Degradation Processes in Reliability

Mathematical Models and Methods in Reliability Set

coordinated by
Nikolaos Limnios and Henry Bo Lindqvist

Volume 3

Degradation Processes in Reliability

Waltraud Kahle
Sophie Mercier
Christian Paroissin

ISTE

WILEY

First published 2016 in Great Britain and the United States by ISTE Ltd and John Wiley & Sons, Inc.

ISTE Ltd
27-37 St George's Road
London SW19 4EU
UK

www.iste.co.uk

John Wiley & Sons, Inc.
111 River Street
Hoboken, NJ 07030
USA

www.wiley.com

Library of Congress Control Number: 2016936598

British Library Cataloguing-in-Publication Data
A CIP record for this book is available from the British Library
ISBN 978-1-84821-888-8

Contents

Abbreviations and Notations ix

Introduction xi

Chapter 1. Wiener Processes 1

1.1. Gaussian distribution 1
1.2. Brownian motion 3
1.3. Constructions and simulation methods 4
1.3.1. Random walk approximation 5
1.3.2. Brownian bridge sampling 6
1.3.3. Karhunen–Loève approximation 9
1.4. Wiener process (Brownian motion with drift) 10
1.5. First passage time (time to failure) 12
1.6. Statistical inference 20
1.6.1. Statistical inference for degradation data 22
1.6.2. Statistical inference for time to failure data 30
1.6.3. Statistical inference for both degradation and time to failure data 31
1.7. Extension and related models 45
1.7.1. Time-scaled Wiener processes 45
1.7.2. Brownian motion with nonlinear drift 46

Chapter 2. Gamma Processes 49

2.1. Gamma distribution 49
2.2. Poisson random measures 54
2.3. Definition and basic properties of a gamma process 68
2.4. Series representations 71

2.5. Simulation 81
2.5.1. Random walk approximation 81
2.5.2. Gamma bridge sampling 84
2.5.3. Compound Poisson process approximations 85
2.5.4. Numerical experiments 89
2.6. Hitting time (time-to-failure) and overshoot 97
2.6.1. Introduction 97
2.6.2. Theoretical results 98
2.6.3. Simulation 109
2.7. Statistical inference in parametric models 111
2.7.1. Estimation based only on degradation data 113
2.7.2. Other cases 124
2.8. Statistical inference in the semi-parametric model 126
2.8.1. Approach based on the pseudo-likelihood function 126
2.8.2. Approach based on an EM algorithm 129
2.9. Statistical inference for models with random effects (frailty model) 134
2.9.1. Parametric model 134
2.9.2. Semi-parametric model 138
2.10. Extensions and related models 145
2.10.1. Accelerated gamma process 145
2.10.2. Perturbed gamma process 147
2.10.3. Extensions of the gamma process 147

Chapter 3. Doubly Stochastic Marked Poisson Processes . . 149

3.1. Modeling degradation by marked point processes 150
3.1.1. The sequence $(T_n)_{n\geq 1}$ 151
3.1.2. Marking of the sequence (T_n) 157
3.1.3. The sequence (Z_n) and the degradation process (X_t) 158
3.1.4. The first passage time τ_h 163
3.2. Statistical inference 167
3.2.1. Maximum likelihood estimations 167
3.2.2. Confidence estimations 176
3.2.3. Moment estimates 179
3.2.4. Comparison of maximum likelihood and moment estimates 182

Chapter 4. Model Selection and Application to Real Data Sets 185

4.1. Takeda device data 185
4.2. Laser data 187
4.2.1. Parameter estimation 189
4.2.2. Comparison of the estimations for the average degradation and the time to failure 189
4.2.3. Procedures for selection model 190
4.3. Hudak crack growth data 195

Bibliography 199

Index 209

Abbreviations and Notations

Abbreviations

càdlàg	"continu à droite avec des limites à gauche" (right-continuous with left-side limits)
c.d.f.	cumulative distribution function
c.p.u. time	central processing unit time (processing time)
i.i.d.	independent and identically distributed
p.d.f.	probability distribution function
r.v.	random variable
IFR	increasing failure rate
MLE	maximum likelihood estimator
MME	moments method estimator

Notations

$\mathbb{N}$	set of non-negative integers
$\mathbb{R}$	set of real numbers
$\mathbb{R}_+$	set of non-negative real numbers
$\mathbb{R}_+^*$	set of positive real numbers
$\bar{\mathbb{R}}_+, \bar{\mathbb{R}}_+^*, \bar{\mathbb{N}}, \cdots$	$\mathbb{R}_+ \cup \{+\infty\}, \mathbb{R}_+^* \cup \{+\infty\}, \mathbb{N} \cup \{+\infty\}, \cdots$
$\leq (\geq)$	smaller (larger) than or equal to
$(a, +\infty)$	open half line
$[a, +\infty)$	closed half line
$*$	convolution product
$X = (X_t)_{t \geq 0}$	stochastic process

$X^{(i)} = \left(X_t^{(i)}\right)_{t \geq 0}$	i-th copy of the stochastic process X
F_U	cumulative distribution function of the r.v. U
f_U	probability distribution function of the r.v. U
$\bar{F}_U$	survival function of the r.v. U
F_t	cumulative distribution function of the r.v. X_t
f_t	probability distribution function of the r.v. X_t
$\bar{F}_t$	survival function of the r.v. X_t
μ_X	Lévy measure of the Lévy process X
$\mathbf{1}_A$	indicator function of the set A
τ_A	first passage time in the set A
τ_h	first passage time in the set $A = (h, +\infty)$
$\mathbb{P}$	probability
$\mathbb{P}_x$	conditional probability given that the initial state is x
$\mathbb{E}$	expectation with respect of $\mathbb{P}$
$\mathbb{E}_x$	expectation with respect of $\mathbb{P}_x$
$\text{Var}(U)$	variance of the r.v. U
$\text{Cov}(U, V)$	covariance of the r.v. U and V
$\mathcal{L}_U$	Laplace transform of the non-negative r.v. U
$\mathcal{L}_t$	Laplace transform of the non-negative r.v. X_t
$\mathcal{G}(a, b)$	gamma distribution with parameters (a, b)
$\mathcal{E}(\lambda)$	exponential distribution with mean $1/\lambda$
$\mathcal{B}(a, b)$	beta distribution with parameters (a, b)
$\mathcal{P}(\lambda)$	Poisson distribution with mean λ
$\mathcal{U}([a, b])$	Uniform distribution on $[a, b]$, with $a < b$
δ_a	Dirac mass at point a
$(\Omega, \mathcal{A}, \mathbb{P})$	probability space, where $\mathcal{A}$ is a σ-algebra on Ω and $\mathbb{P}$ a probability measure on $(\Omega, \mathcal{A})$
$\mu \otimes \nu$	product measure of the two measures μ and ν
$\mathcal{B}(E)$	Borel σ-algebra on E (which may be $\mathbb{R}$, $\mathbb{R}^d$, $\mathbb{R}_+$, etc.)
$\mathcal{N}(\mu, \sigma^2)$	one-dimensional Gaussian distribution
$\mathcal{N}(\bar{\mu}, \Sigma)$	n-dimensional Gaussian distribution
$\stackrel{d}{=}$	equal in distribution
$B = (B_t)_{t \geq 0}$	Brownian motion
$W = (W_t)_{t \geq 0}$	Wiener process

Introduction

The terminology degradation process refers to many types of models in reliability, which correspond to various kinds of stochastic processes used for deterioration modeling. This book is restricted to the case of a continuous set of possible deterioration levels, to be opposed for instance to multi-state models, where only a finite number of deterioration levels is possible (see, for instance, [COM 99, HOU 99, AND 02] for reviews on these multi-state models with application in epidemiology). Also, only univariate models are considered, which means that one single measurable quantity is assumed to be observed over time. In this book, the terms degradation model or degradation process refer to this context.

Even within this apparently restrictive context, degradation models have been much studied for several decades. Nowadays, there exists a large number of publications on this topic in scientific journals. However, only a few books deal with it. Some monographs contain one single chapter on degradation data and cannot cover a wide range of degradation models. For instance, Chapter 13 in [MEE 98] deals with the degradation path model (see section I.3 below for details) while Chapter 11 in [NEL 02] and Chapter 3 in [BAG 04] focus on some accelerated degradation models. One can however quote [FIN 13], which is mostly devoted to shocks and burn-in models and does not cover the range of this book. Beside these monographs, some books gather some material from conferences in mathematical reliability (see [NIK 10] or [RYK 10], for instance). They clearly contain interesting papers on degradation models but they do not share the same unity as a monograph and they do not allow us either to go as deeply into these models. Other books exist that mostly focus on the study of maintenance policies and their optimization, see [NAG 07].

Consequently, a monograph uniquely devoted to this topic is lacking. This book aims at filling this gap. Well-known results together with new results are provided here in a unique rigorous mathematical framework. We hope that it will be useful for researchers and PhD students in applied mathematics, but also for research engineers dealing with degradation modeling and maintenance. It can also be used as a basis for teaching applied stochastic processes.

After this brief overview, we now go on with the core of this introduction. We first explain why it is interesting to study degradation models. A few real data sets are presented in the following for illustrative purposes. The most classical degradation models are then introduced, which are nearly all based on specific Lévy processes or on some of their extensions. Basic definitions and properties are next provided for such mathematical objects. Finally, an outline of the book is sketched in the last section.

I.1. From lifetime data to degradation path observations

Historically, the first reliability studies [BAR 65, BAR 75] were focused on lifetime data analysis, where equipment (or component) lifetimes were directly modeled through random variables. These studies belong to the so-called survival analysis that had (and still has) a major role in reliability engineering [MEE 98] but also in many other fields such as biology and health [KLE 03], demography, actuarial science, economics and social sciences (see, for instance, the bibliographical notes of Chapter 1 in [LAW 03] for further references).

However, purely lifetime-based reliability analysis is not always an effective means to analyze the available data. First, samples of lifetime data can be very small or highly censored. Indeed, observing an item up to failure can be very long and costly, especially for highly reliable products. For some applications, lifetime data are even not observed at all due to the severity of the failure (such as in nuclear engineering, for instance). Second, the development of on-line monitoring together with the increasing use of sensors for safety assessment now allow us to have more and more available information. These data collected on-line can correspond to some covariate process, which has some influence on the lifetime data. Most of the time, they are however the reflection of some deterioration (or degradation) mechanism, which cannot be handled with lifetime models. This includes, for instance, processes of wear, corrosion, crack growth, vibration, etc. Often, this degradation is measured through a scalar indicator, which tends to increase over time. This scalar indicator is hereafter called degradation measurement. In most cases, there exists a prescribed critical level above which a piece of equipment is considered not to be able to consistently perform its intended mission any more, either for safety or technical reasons. The time required to reach this critical level corresponds to the equipment lifetime, which of course remains of the first interest, even considering degradation data. However, the degradation data bring some additional information on the equipment status when compared to dealing only with failure data. The advantage of considering degradation data toward lifetime data has, for instance, been highlighted from an example by Meeker and Escobar [MEE 98, page 335].

From a probabilistic point of view, although lifetime data stand for a sample of non-negative random variables, successive degradation data represent observations of a stochastic process, whose parameters (eventually functional) can be fit from the data. A main interest is that once fitted, it next allows us to make some prediction over the future of the equipment. As an example, given its present deterioration level, the distribution of its (future) residual life can be derived (as well as many other quantities of interest in an industrial context). This is of major importance to define complex and powerful preventive maintenance policies that can now be based on the effective status of the equipment and on its known future (random) evolution. See [VAN 09] for a large survey on maintenance policies devoted to the gamma process (a specific degradation process studied in Chapter 2), with also some references on other processes such as the Wiener process (studied in Chapter 1). See also [AHM 12] and [JAR 06] for an overview on condition-based maintenance policies in a more general setting. Beyond the definition of efficient maintenance policies, a last interest of degradation models is that they can be used for test planning and burn-in modeling, see sections 5 and 6 in [YE 15] for a survey on these two topics as well as [FIN 13] for the second topic.

For a better understanding of which kind of data sets we are thinking about, we now introduce a few real data sets from the literature, which will be analyzed in the final chapter of the book, with the tools developed therein.

I.2. A few real data sets

The first data set is called Takeda device data in the literature and it has been first studied by Takeda and Suzuki [TAK 83] (see also [STI 89] or [LU 97], for instance). These authors have measured the performances of a certain kind of semiconductors, called metallized and oxidized semiconductors field-effect transistors. Such semiconductors are subjected to one particular type of degradation, namely, the hot carrier degradation. For more details, the reader could refer to the monograph by Leblebici and Kang [LEB 93], especially Chapters 1 and 2. The critical level is fixed at 15%. Five specimens were observed at 35 instants, giving a total of 170 increments. All the observation times (and thus, in particular, the first ones) are the same for all units. The data set contains positive (126), null (32) and negative (12) increments. The data has been reproduced in Figure I.1 in a log-log scale, together with the critical level (dot line). The critical level is assumed to be equal to 15% of transconductance degradation [LU 97]. Clearly, after a log-log transformation, degradation appears to be linearly increasing with respect to the time in average. We can notice that three units among the five specimens are unfailed at the end of the experience. Hence, an interesting problem is the estimation of the remaining lifetime distribution.

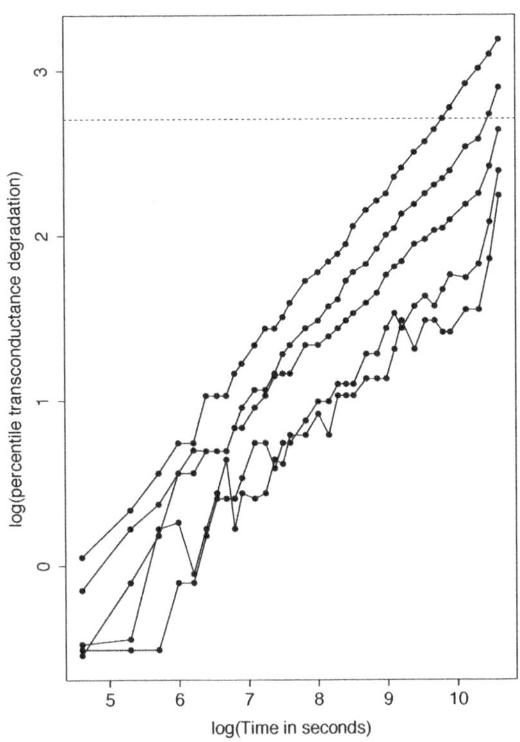

Figure I.1. *Takeda device data*

The second data set has been introduced by Meeker and Escobar [MEE 98], known in the literature as laser data. A set of 15 gallium arsenide laser devices is tested at 80° and periodically observed. More precisely, the performance characteristics of the devices are observed each 250 h until 4,000 h (observation times are identical for all units). A device is considered to be failed when its performance characteristic reaches the predefined critical level equal to 10. The data are plotted in Figure I.2. For this data set, we will be able to fit two different degradation parametric models. Hence, naturally, the problem of selection model arises: which model seems to be more suitable? After estimating the parameters for these two models, we will offer a discussion about this problem.

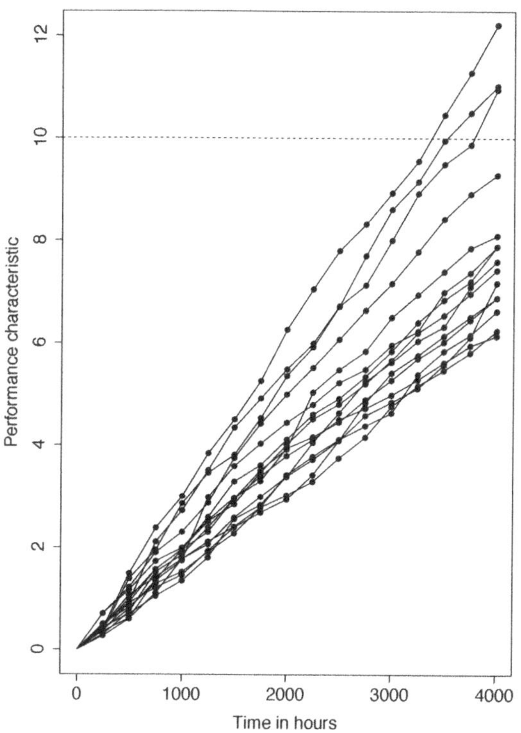

Figure I.2. *Laser data*

The third and last data set deal with crack growth. Hudak *et al.* [HUD 78] have observed 21 metallic specimens, each subjected to loading cycles with crack length recorded every 10^4 cycles. Initial crack length was 0.9 inches for each specimen. The average degradation is clearly nonlinear *a contrario* to the previous data set. The data are plotted in Figure I.3. Such a behavior is typical for crack growth (see, for instance, another famous data which is the Virkler data, [VIR 79]). Hence, the choice of a parametric degradation model is not obvious. We will see that, for one degradation model, semi-parametric estimators are available. Hence, we will be able to compare parametric models to the semi-parametric model.

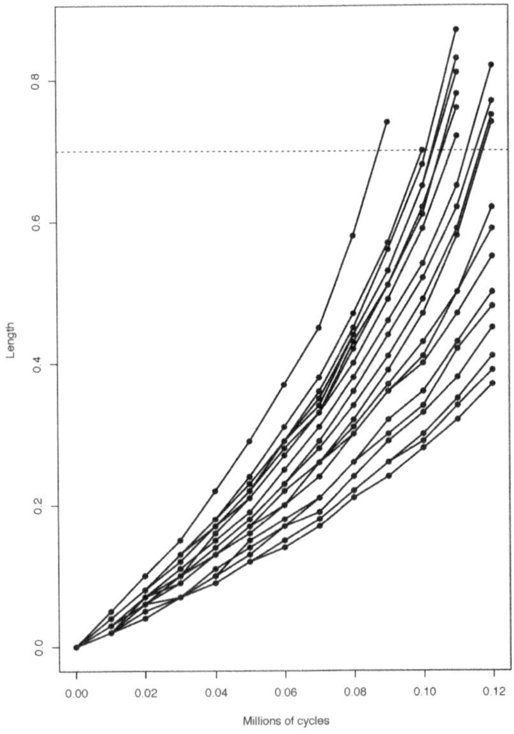

Figure I.3. *Hudak crack growth data*

I.3. A brief overview of classical degradation models

As already mentioned, this book is devoted to the case of a univariate degradation with a continuous set of possible outcomes. The degradation process is assumed to take range in an interval or half-line of $\mathbb{R}$, or in $\mathbb{R}$ itself. Hence, only this case is considered in this brief overview.

A first approach for modeling this kind of deterioration relies on the assumption that the degradation is a monotone and deterministic phenomenon, but that there exists a unit-to-unit variability, which can be translated through some random coefficients. More specifically, the assumption is that the degradation X_{ij} of unit i at time t_{ij} is given by

$$X_{ij} = \mathcal{D}(t_{ij}, \beta_{1i}, \dots \beta_{ki}) + \epsilon_{ij}, \quad i \in \{1, \dots, n\}, \quad j \in \{1, \dots, m_i\},$$

where $\mathcal{D}$ is a deterministic function, ϵ_{ij} is Gaussian distributed with mean 0 and variance σ_ϵ^2, and β_{li}'s are random parameters. The model is called a degradation path model, see [MEE 98, Chapter 13] for a comprehensive introduction to this approach. The β_{li}'s can differ from one unit to the other or (some of them) can be common across all units. Degradation path models are very similar to well-known generalized linear models with fixed and random effects, which have already been the object of several books, such as [MCC 01] for instance. We consequently do not consider them in the present book. In a degradation path model, given the realization of the β_{li}'s for one specific item, all the randomness is contained in the ϵ_{ij}'s, which can be seen as measurement errors. These models are useful when a deterministic description of the degradation is available, based on mechanical or physical considerations. When it is not the case, other models are required.

A second approach for modeling deterioration (with continuous range) is based on the use of specific Lévy processes and on some of their extensions. This approach seems to date from the middle of 1970s (see the survey by van Noortwijk [VAN 09]). The most classical Lévy processes used for deterioration modeling are the Wiener and the gamma processes, which are dealt with in Chapters 1 and 2 of this book, respectively. Note that these two processes are often used in their non-homogeneous versions, which are not Lévy any more but direct extensions. To our knowledge, Abdel-Hameed was the first to consider the gamma process as a wear model in 1975 [ABD 75]. Çinlar and co-authors justify its use for extrapolating concrete creep in 1977, based on physics considerations and on the mathematical properties of the process. From an applicative perspective, the gamma process is well suited to model some non-decreasing deterioration, which accumulates over time through many tiny increments. Indeed, it is a pure jump process with an infinitely number of jumps over any finite time interval (see Chapter 2 for details). On the contrary, a Wiener process models a continuous deterioration with an increasing trend but a non-monotonous trajectory (see Chapter 1). It seems that the use of the Wiener process for deterioration modeling started around the 1990s, see for instance [DOK 92] or [DOK 95]. During the last two decades, both gamma and Wiener processes has received much consideration in the literature. These two models correspond to some mostly continuous deterioration (which is the fruit of an accumulation of a large amount of tiny increments for the gamma process). However, the deterioration can also be due to isolated shocks on a system, where each shock induces a random amount of damage that is accumulating over time. In that case, the trajectory of the deterioration is piecewise constant and the two previous models are not adapted.

In the case of a piecewise constant deterioration trajectory, the classical approach is to consider a cumulative shock model, such as the compound-renewal damage model introduced by Morey in 1966 [MOR 66]. In the specific case where the renewal process is a homogeneous Poisson process, we obtain a compound Poisson process, which is a Lévy process again. There are a lot of papers dealing with shock models under various assumptions, such as [ÇIN 77a, ESA 73, FEN 94, SHA 83, SOB 87] and we do not intend to review all the possible models here, neither to describe them all in the book. We concentrate in Chapter 3 on the specific case where the arrival process for the shocks is a doubled-stochastic Poisson process, namely, where the intensity of the Poisson process is random. This process was introduced by Cox [COX 55]. The random amounts of damage are often considered as independent from their arrival times in the literature. This assumption is relaxed in Chapter 3, which hence appears to be dealt in a quite general setting.

I.4. The mathematical framework

All the degradation models studied in the book are stochastic processes. We here remind well-known facts on this notion and provide some of our notations. All of the stochastic processes considered hereafter take range in $\mathbb{R}^d$ (or $\mathbb{R}_+^d$), where $d \in \mathbb{N}^*$ and we set $\mathcal{B}(\mathbb{R}^d)$ to stand for the Borel σ-algebra on $\mathbb{R}^d$. Let $(\Omega, \mathcal{A})$ be a measurable space and let $T = \mathbb{R}^+$ or $T = \mathbb{N}$. Following [SAT 99], we call stochastic process with state space $(\mathbb{R}^d, \mathcal{B}(\mathbb{R}^d))$ and parameter set T, any collection $X = (X_t)_{t\in T}$, where, for each $t \in T$, X_t

is a random variable defined on $(\Omega, \mathcal{A})$ which takes values in $(\mathbb{R}^d, \mathcal{B}(\mathbb{R}^d))$. The function $\omega \to X(\omega) = (X_t(\omega))_{t\in T}$ is then a random variable from $(\Omega, \mathcal{A})$ to $(E^T, \mathcal{E}^T)$, where E^T stands for the set of all functions from T to $E = \mathbb{R}^d$ and $\mathcal{E}^T$ is the σ-field generated by all finite sets of such functions.

Now, let us recall that a stochastic process X is said to be càdlàg ("*continu à droite avec des limites à gauche*" in French, namely, right-continuous with left-side limits), if the trajectory $t \to X_t(\omega)$ is right-continuous with left-side limits, for each $\omega \in \Omega$. Let us equip the measurable space $(\Omega, \mathcal{A})$ with a probability measure $\mathbb{P}$ so that $(\Omega, \mathcal{A}, \mathbb{P})$ becomes a probability space. Then, it is known that each stochastic process X on $(\Omega, \mathcal{A}, \mathbb{P})$ with state space $(\mathbb{R}^d, \mathcal{B}(\mathbb{R}^d))$ admits a càdlàg modification, which means that there exits a càdlàg process Y such that

$$\mathbb{P}\left(\omega \in \Omega : X_t\left(\omega\right) = Y_t\left(\omega\right) \text{ for all } t \in T\right) = 1.$$

As a result, there is no loss of generality assuming the process X to be càdlàg, which is done from now on and throughout the book.

In the specific context of deterioration modeling, additive accumulation of degradation and independence of degradation increments at disjoint time intervals are natural assumptions. These properties lead immediately to the conclusion that a suitable class of processes for degradation modeling are additive processes and among them, Lévy processes, see [SAT 99] for more details on these processes and for the following definitions.

DEFINITION I.1.– *Let $X = (X_t)_{t\geq 0}$ be a (càdlàg) stochastic process from $(\Omega, \mathcal{A}, \mathbb{P})$ to $(\mathbb{R}^d, \mathcal{B}(\mathbb{R}^d))$. Then, X is said to be an additive process if the following conditions are satisfied:*

1) $X_0 = 0$ *$\mathbb{P}$-a.s.;*

2) *X has independent increments, i.e. for all $n \in \mathbb{N}^*$, for all $t_1 < \cdots < t_n$, the random variables $X_{t_1}, X_{t_2} - X_{t_1}, \cdots, X_{t_n} - X_{t_{n-1}}$ are independent;*

3) *X is stochastically continuous, i.e. for any $s, t \geq 0$ and for any $\varepsilon > 0$,* $\lim_{t\to s} \mathbb{P}\left(|X_s - X_t| > \varepsilon\right) = 0$.

The stochastic process X is said to have stationary increments, if, for any $s, t \geq 0$, the distribution of $X_{t+s} - X_t$ does not depend on t. An additive process with stationary increments is called a Lévy process.

Note that additive processes are also called non-homogeneous Lévy processes in the literature. Also, if $X = (X_t)_{t\geq 0}$ is a Lévy process, then both mean and covariance matrix of X_t are linear functions of the time t (whenever they exist). Finally, starting from a Lévy process X and an increasing function $A : \mathbb{R}_+ \longrightarrow \mathbb{R}_+$ such that $A(0) = 0$, then $\left(Y_t = X_{A(t)}\right)_{t\geq 0}$ is an additive process, see [SAT 99, example 1.7].

Now, let us consider instants $0 = t_0 < t_1 < \cdots < t_n$ and, for any $i \in \{1, \ldots, n\}$, let X_{t_i} stand for the deterioration level at time t_i and let $Y_i = X_{t_i} - X_{t_{i-1}}$ be the degradation increment between t_{i-1} and t_i. Based on the additive assumption of the deterioration, we should have $X_{t_n} = Y_1 + \cdots + Y_n$ (roughly speaking, this means that there is no conflict between measurements of deterioration at different time points). A necessary condition is that the distribution of X_{t_n} should be the same as the distribution of $Y_1 + \cdots + Y_n$. Assuming furthermore the increments to be independent, this mostly allows us to the requirement that the distribution of X_t should be infinitely divisible.

DEFINITION I.2.– *The distribution $\mathbb{P}_X$ of a random variable X is infinitely divisible if, for any $n \in \mathbb{N}^*$, there exists i.i.d. random variables $X_1^{(n)}, \ldots, X_n^{(n)}$ such that*

$$X \stackrel{d}{=} X_1^{(n)} + \cdots + X_n^{(n)}$$

where $\stackrel{d}{=}$ stands for equality in distribution.

It can be proved [SAT 99, theorem 9.1] that for any additive process $X = (X_t)_{t\geq 0}$ (and hence for any Lévy process), the distribution of X_t is infinitely divisible, for all $t \geq 0$. Examples of Lévy processes are the Wiener and the homogeneous gamma process described in Chapters 1 and 2. Gaussian and gamma distributions are consequently infinitely divisible. Examples of additive processes are the non-homogeneous gamma process described in Chapter 2 and the non-homogeneous compound Poisson process, which is a specific case of the doubly stochastic marked Poisson process considered in Chapter 3.

I.5. Outline of the book

The book contains three main chapters, each devoted to one degradation model. The last chapter contains some illustrative studies, based on the previously described real data sets. For each degradation model, we begin by providing probabilistic results (moments, time-to-failure distributions, etc.) and exploring simulation tools for sample path generation. Next, various estimation procedures are developed, depending on the available data (degradation measurements and/or failure times). In each case, both maximum likelihood and moments methods are considered and asymptotic properties provided.

In Chapter 1, the Wiener process is considered as a candidate for a degradation model. After some recall on Gaussian random vectors, the Brownian motion is introduced and its construction provided, together with different simulation methods. Next, the Wiener process is defined and its first-passage time distribution is studied. For this model, the latter is available in full form and known as the inverse Gaussian distribution. Statistical inference is next considered for various types of sampling schemes. Since the Wiener process has a time-linear expectation (as any Lévy process), it is not always a suitable approach for deterioration modeling. Hence, to conclude this chapter, we provide a short review of some extensions of the Wiener process and some related models, which are not Lévy processes any more.

Chapter 2 is devoted to the gamma process, which is considered in his most general non-homogeneous version. We first recall basic facts on the gamma distribution. For a better understanding of the jump structure of a gamma process, we next introduce Poisson random measures on $\mathbb{R}^d$, from where gamma processes are next constructed. The section on Poisson random measures is intended to be self-contained since it is not necessarily part of the usual basic knowledge of reliability researchers. It can be skipped in a first reading. We next detail the presentation of the gamma process and provide several simulation tools, whose efficiency is studied both theoretically and numerically. Time-to-failures are next investigated from a probabilistic point of view and simulation tools are numerically studied. Just as for the Wiener process, statistical inference is next considered for various sampling schemes. Random effects (or frailty) are also envisioned, which allows us to introduce some variability between individuals. Finally, a brief review is proposed about extensions of the gamma process and related models.

In Chapter 3, a cumulative shock model is studied, where the shocks arrive according to a doubly stochastic Poisson process (or Cox process), with intensity assumed to be a random multiple of a deterministic function. Just as for the gamma process, this allows us to introduce some variability between individuals. Each shock induces an instantaneous increment of the degradation level, which is allowed to depend on the time of the shock, leading to a so-called position-dependent marking of the doubly stochastic Poisson process. Different parametric models are envisioned, for which probabilistic results are provided (including on the first passage time distribution) and estimation procedures developed.

Finally, Chapter 4 illustrates the use of the previously described degradation models on the real data sets from section I.2 (except however for the last model from Chapter 3, for which we had no real data at our disposal). This chapter allows us to highlight some differences between the Wiener and the gamma processes for modeling purposes and to illustrate the use of the previously developed probabilistic and statistical tools. Some procedures for selecting a model and testing its goodness of fit are also proposed in the last chapter.

1

Wiener Processes

The Wiener process is one of the easiest models for random accumulation of degradation over time. It is based on the assumption of an additive accumulation of degradation with linear wear intensity. Regarding every degradation increment as an additive superposition of a large number of small effects, we can assume the degradation process to be normally distributed.

1.1. Gaussian distribution

Since the Wiener process is based on normally distributed increments, we recall here some basic properties of the Gaussian (normal) distribution.

DEFINITION 1.1.– *A random variable X is said to be (univariate) Gaussian distributed, or normally distributed, with mean $\mu \in \mathbb{R}$ and variance $\sigma^2 > 0$ if its distribution admits the following probability density function (p.d.f.) with respect to Lebesgue measure:*

$$\forall x \in \mathbb{R}, \quad f_X(x) = \frac{1}{\sqrt{2\pi\sigma^2}} \exp\left\{-\frac{(x-\mu)^2}{2\sigma^2}\right\}.$$

By extension, a constant random variable $X = \mu$ with $\mu \in \mathbb{R}$ is said to be Gaussian distributed with mean μ and variance $\sigma^2 = 0$. In any case, we write $X \sim \mathcal{N}(\mu, \sigma^2)$.

The specific case $\mathcal{N}(0, 1)$ where $\mu = 0$ and $\sigma = 1$ is called the standard Gaussian (normal) distribution.

DEFINITION 1.2.– *An n-dimensional random vector $X = (X_1, \ldots, X_n)^T$ is said to be (multivariate) Gaussian distributed (or normally distributed or* *a Gaussian vector) if any linear combination of its components is a normally distributed random variable:*

$$\forall a_1, \ldots, a_n \in \mathbb{R}, \quad a_1 X_1 + \cdots + a_n X_n$$

is (univariate) normally distributed.

A direct consequence of the definition is that Gaussian distributions are stable under marginalization: Any subvector of a Gaussian vector still is a Gaussian vector. In particular, all the components of a Gaussian vector are Gaussian distributed random variables. Also, any random vector that is obtained through affine transformation of a Gaussian vector is a Gaussian vector. Furthermore, using the Fourier transform, it can be checked that the distribution of a Gaussian vector X only depends on its mean vector μ and covariance matrix Σ. The corresponding distribution is denoted by $\mathcal{N}_n(\mu, \Sigma)$, where n stands for the dimension of X. We now get the following result, where the symbol ^{T} means transposition.

PROPOSITION 1.3.– Let $X = (X_1, \ldots, X_n)^T$ be a Gaussian vector with $X \sim \mathcal{N}_n(\mu, \Sigma)$. Let A be a $p \times n$ matrix and B a $p \times 1$ column vector. Then, $AX + B$ is a Gaussian vector with distribution $\mathcal{N}_p(A\,\mu + B, A\,\Sigma\,A^T)$.

We can also check that gathering independent Gaussian random variables $X_1, \ldots, X_n$ provides a Gaussian vector $(X_1, \ldots, X_n)$. As a specific case, gathering n independent and identically distributed (i.i.d.) standard Gaussian random variables provides a standard Gaussian vector with distribution $\mathcal{N}_n(0_n, I_n)$, where 0_n refers to an n-dimensional vector and I_n to the n-dimensional identity matrix.

The previous results provide a way to construct a random vector with a specific Gaussian distribution.

COROLLARY 1.4.– Let Y be a n-dimensional standard Gaussian vector ($Y \sim \mathcal{N}_n(0, I_n)$). Let Σ be a symmetric and positive semi-definite $n \times n$ matrix and μ be an n-dimensional vector. Let A be a matrix such that $\Sigma = AA^T$. Then, the random vector $X = AY + \mu$ is a Gaussian vector with mean μ and covariance matrix Σ.

Assuming further that Σ is non-singular (and hence symmetric positive definite), it is easy to derive from the previous corollary that a Gaussian vector $X \sim \mathcal{N}_n(\mu, \Sigma)$ admits the following p.d.f. with respect to Lebesgue measure:

$$\forall x \in \mathbb{R}^n, \quad f_X(x) = \frac{1}{\sqrt{(2\pi)^n \det \Sigma}} \exp\left\{-\frac{1}{2}(x-\mu)^T\,\Sigma^{-1}(x-\mu)\right\},$$

where $x = (x_1, \ldots, x_n)^T$, Σ^{-1} is the inverse matrix of Σ and $\det \Sigma = |\Sigma|$ is its determinant.

Whether Σ is non-singular or not, an important property of a Gaussian vector is that its components are independent random variables as soon as the covariance matrix Σ is diagonal. As mentioned in the Introduction, the Gaussian distribution is also infinitely divisible.

All the processes that we consider in the rest of this chapter are Gaussian processes, which we now define.

DEFINITION 1.5.– *A time-continuous stochastic process $X = (X_t)_{t \geq 0}$ is said to be a Gaussian process if, for any $n \in \mathbb{N}^*$ and for any $(t_1, \ldots, t_n) \in \mathbb{R}_+^n$, the random vector $(X_{t_1}, \ldots, X_{t_n})$ is a Gaussian vector.*

1.2. Brownian motion

We here introduce and provide basic properties of the well-known Brownian motion, which is a specific Wiener process and from where general Wiener processes will next be constructed.

DEFINITION 1.6.– *$B = (B_t)_{t \geq 0}$ is a Brownian motion if, and only if,*

1) *$B_0 = 0$ a.s.;*

2) *B has stationary increments: for any $t, s, h > 0$, $B_t - B_s \overset{d}{=} B_{t+h} - B_{s+h}$;*

3) *B has independent increments: for any $n \geq 1$ and for any $t_1 < \cdots < t_n$, the random variables $B_{t_1}, B_{t_2} - B_{t_1}, \ldots, B_{t_n} - B_{t_{n-1}}$ are independent;*

4) *for any $t > 0$, $B_t \sim \mathcal{N}(0, t)$.*

Figure 1.1 shows some simulations of Brownian motion paths.

The Brownian motion has a number of remarkable properties:

1) the Brownian motion is a Gaussian process. In particular, $(B_{t_1}, \ldots B_{t_n}) \sim \mathcal{N}_n(\mu, \Sigma)$ with mean $\mu = 0_n$ and covariance matrix $\Sigma = (\min\{t_i, t_j\})_{i,j \in \{1,\ldots,n\}^2}$;

2) the Brownian motion is a Lévy process and it is the only Lévy process with a.s. continuous sample paths;

3) the sample paths are almost surely not differentiable anywhere and have unbounded variation;

4) reflection property. If the path of a Brownian motion reaches a value a at time $t = s$, then the subsequent path after time s is identically distributed as the horizontally reflected path, starting from point a at time s;

5) self-similarity. For $0 < t_1 < t_2 < \cdots < t_n$ and $T > 0$

$$(\sqrt{T}B_{t_1}, \sqrt{T}B_{t_2}, \ldots, \sqrt{T}B_{t_n}) \stackrel{d}{=} (B_{Tt_1}, B_{Tt_2}, \ldots, B_{Tt_n}).$$

For an extensive description of Brownian motion and its properties, the reader could refer to [MÖR 10], for instance.

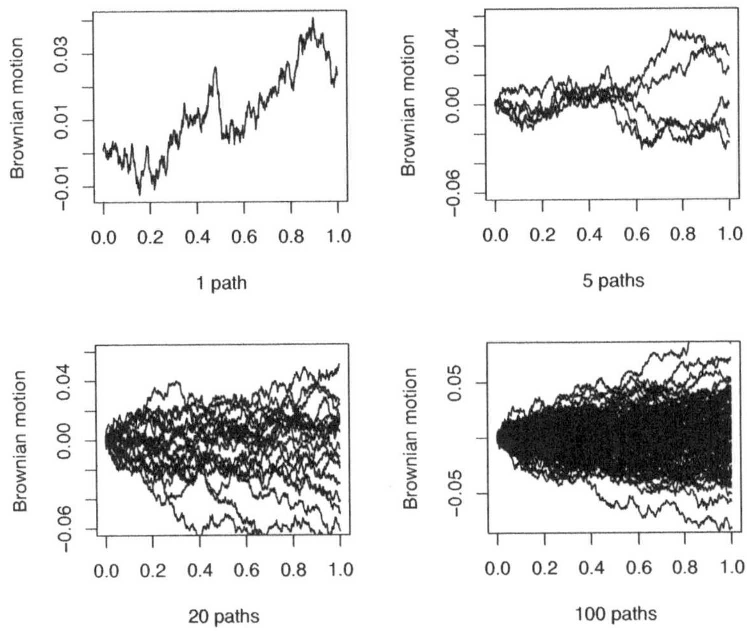

Figure 1.1. *Simulations of Brownian motion path*

1.3. Constructions and simulation methods

In this section, we provide three constructions of the Brownian motion, leading to different simulation methods.

1.3.1. *Random walk approximation*

Let $(X_n)_{n\geq 1}$ be a sequence of i.i.d. random variables with zero mean and unit variance. We consider the random walk $(R_n)_{n\geq 0}$ defined by:

$$R_0 = 0 \quad \text{and} \quad R_n = X_1 + \cdots + X_n, \quad n \geq 1.$$

Then, by the central limit theorem (see theorem 4.4, page 120 in [PET 95], for instance),

$$\sup_{x\in\mathbb{R}} \left| \mathbb{P}\left(\frac{R_n - \mathbb{E}R_n}{\sqrt{\mathrm{Var}(R_n)}} \leq x \right) - \Phi(x) \right| \xrightarrow[n\to\infty]{} 0\,,$$

where Φ is the cumulative distribution function (c.d.f.) of the standard Gaussian distribution (i.e. with zero mean and unit variance). Consider the process with continuous sample paths on $[0,1]$: [MIK 98]

$$S_n(t) = \begin{cases} R_i/\sqrt{n} & \text{for } t = \frac{i}{n} \text{ with } i \in \{1,\ldots,n\} \\ \text{linearly interpolated elsewhere.} \end{cases}$$

Let us consider, for the moment, the restriction of the process S_n at points i/n and let us assume that the distribution of $X_1, X_2, \ldots$ is the standard Gaussian distribution. Then, the following properties hold:

– $\mathbb{P}(S_n(0) = 0) = 1$;

– $(S_n(t))_{t\geq 0}$ has independent increments in the following sense: for any $(i_1,\ldots,i_m) \in \mathbb{N}^m$ with $m < n$ and $0 \leq i_1 < \ldots < i_m = n$,

$$S_n(i_2/n) - S_n(i_1/n), \ldots,\ S_n(i_m/n) - S_n(i_{m-1}/n)$$

are independent;

– for all $0 \leq i \leq n$, $S_n(i/n) \sim \mathcal{N}(0, i/n)$.

Hence, $S_n(t)$ and B_t for $t \in \{0, 1/n, 2/n, \ldots, 1\}$ share a lot of identical properties.

By the *functional central limit theorem*, also called *Donsker's invariance principle* [BIL 99, theorem 8.2, p. 90], the process $(S_n(t))_{t\geq 0}$ converges uniformly on $C[0,1]$, the space of continuous functions on the unit interval, to the Brownian motion on $[0,1]$. For all possible choices of $t_i \in [0,1]$ and $x_i \in \mathbb{R}$, for $i \in \{1,\ldots,m\}$ with $m \geq 1$,

$$\mathbb{P}(S_n(t_1) \leq x_1, \ldots, S_n(t_m) \leq x_m) \xrightarrow[n\to\infty]{} \mathbb{P}(B_{t_1} \leq x_1, \ldots, B_{t_m} \leq x_m).$$

From the self-similarity property of the Brownian motion, we also know how to obtain an approximation of the sample paths of the Brownian motion on any interval $[0,T]$: simulate one path of $(S_n)_{n\geq 0}$ on $[0,1]$, then scale the time interval by the factor T and the sample path by the factor $\sqrt{T}$. Thus, a first simple tool for simulating Brownian sample paths is given by the following algorithm.

ALGORITHM 1.7.– Simulation of a Brownian motion based on random walk approximation.

For $0 = t_0 < t_1 < \cdots < t_k = T$, generate standard normally distributed random variables $X_1, \ldots,$ X_k and recursively define

$$\begin{aligned} B_{t_1} &= \sqrt{t_1}X_1 \\ B_{t_2} &= B_{t_1} + \sqrt{t_2 - t_1}X_2 = \sqrt{t_1}X_1 + \sqrt{t_2 - t_1}X_2 \\ &\vdots \\ B_{t_k} &= \sum_{i=1}^{k} \sqrt{t_i - t_{i-1}}X_i \end{aligned}$$

The paths shown in Figure 1.1 were simulated by this method.

1.3.2. *Brownian bridge sampling*

The Brownian bridge construction is a way to build a path of the Brownian motion by successively adding finer scale details: to simulate a Brownian motion on $[0,T]$, we first simulate a realization of Z_T. Next, we simulate a realization of $Z_{T/2}$ given Z_T and $Z_0 = 0$. Next, a realization of $Z_{T/4}$ given Z_0 and $Z_{T/2}$ is simulated, together with a realization of $Z_{3T/4}$ given $Z_{T/2}$ and Z_T, and so on. At the n-th step, the interval $[0,T]$ is divided into 2^n intervals of the form:

$$I_{n,k} = \left[\frac{k-1}{2^n}T, \frac{k}{2^n}T\right] = [t_{n,k-1}, t_{n,k}], \quad k \in \{1,\ldots,2^n\}.$$

These intervals are chosen in such a way that going from level n to level $n+1$ divides each interval of the n-th level exactly in half.

The Brownian bridge construction builds an approximated Brownian motion path B_t for $t \in [0, T]$ iteratively and level by level. At level n, the path $B_{n,t}$ is constructed through a piecewise linear approximation. This piecewise linear path is determined by its values $B_{n,t_{n,k}}$ at the endpoints $t_{n,k}$, with $k \in \{0, \ldots, 2^n\}$. The construction ensures that these $B_{n,t_{n,k}}$ have the same joint distribution as the corresponding $B_{t_{n,k}}$ of an exact Brownian motion path. Going from stage n to stage $n+1$, one keeps the values $B_{n,t_{n,k}}$ and introduces a new value at the center of each interval $I_{n,k}$. To simulate these new values, we have to calculate the conditional distribution of B_t at the midpoint given the two boundary values. This is done through the following proposition. Note that a Brownian motion in an interval with given boundary values is called *Brownian bridge*.

PROPOSITION 1.8.– Let $0 < s < u < t$. Then, given $B_s = x$ and $B_t = z$, the random variable B_u is Gaussian distributed with

$$\mathbb{E}(B_u | B_s = x, B_t = z) = x + \frac{z - x}{t - s}(u - s)$$

and

$$\mathrm{Var}(B_u | B_s = x, B_t = z) = \frac{(u - s)(t - u)}{t - s}$$

for all $x, z \in \mathbb{R}$.

PROOF.– Let $f_{u|s,t}(\cdot|x, z)$ be the conditional p.d.f. of B_u given $B_s = x$ and $B_t = z$, with

$$\forall y \in \mathbb{R}, \quad f_{u|s,t}(y|x, z) = \frac{f_{s,u,t}(x, y, z)}{f_{s,t}(x, z)},$$

where $f_{s,u,t}$ and $f_{s,t}$ stand for the p.d.f. of (B_s, B_u, B_t) and (B_s, B_t), respectively.

Because of the independent increments of the Brownian motion, we have:

$$f_{s,u,t}(x, y, z) = f_s(x) f_{u-s}(y - x) f_{t-u}(z - y)$$

and

$$f_{s,t}(x, y, z) = f_s(x) f_{t-s}(z - x),$$

where, for all $t > 0$, f_t stands for the p.d.f. of the Gaussian distribution with zero mean and variance t, with

$$\forall x \in \mathbb{R}, \quad f_t(x) = \frac{1}{\sqrt{2\pi t}} \exp\left(-\frac{x^2}{2t}\right).$$

Therefore, the conditional p.d.f. $f_{u|s,t}(y|x,z)$ can be calculated as follows:

$$\begin{aligned} f_{u|s,t}(y|x,z) &= \frac{f_{u-s}(y-x) \cdot f_{t-u}(z-y)}{f_{t-s}(z-x)} \\ &= \frac{1}{\sqrt{2\pi}} \sqrt{\frac{t-s}{(u-s)(t-u)}} \exp\left(-\frac{A}{2}\right) \end{aligned}$$

with

$$\begin{aligned} A &= \frac{(y-x)^2}{u-s} + \frac{(z-y)^2}{t-u} - \frac{(z-x)^2}{t-s} \\ &= \frac{1}{(t-s)(u-s)(t-u)} \left((y-x)(t-u) - (z-y)(u-s)\right)^2 \\ &= \frac{(t-s)}{(u-s)(t-u)} \left(y - x - \frac{z-x}{t-s}(u-s)\right)^2. \end{aligned}$$

It is now easy to recognize that $f_{u|s,t}(\cdot|x,z)$ is the p.d.f. of the Gaussian distribution with the mean and variance given in the proposition. □

For simulation purposes through the Brownian bridge, we need the conditional distribution of B_u given $B_s = x$ and $B_t = z$ at the midpoint, that is for $u = (t+s)/2$. From the previous result, it is easy to see that it is a Gaussian distribution with expectation $(x+z)/2$ and variance $(t-s)/4$. This leads to the following algorithm.

ALGORITHM 1.9.– [Simulation of a Brownian motion based on Brownian bridge sampling] Generate standard normally distributed random variables $X_1, X_2, \ldots$

– Step 0. Set $t_{0,0} = 0$, $t_{0,1} = T$ and $B_{t_{0,0}} = 0$, $B_{t_{0,1}} = \sqrt{T} X_1$.

– Step 1. Leave $B_{t_{1,0}} = B_{t_{0,0}}$, $B_{t_{1,2}} = B_{t_{0,1}}$. Set $\Delta t = T$ and add

$$B_{t_{1,1}} = (B_{t_{1,0}} + B_{t_{1,2}})/2 + \frac{\sqrt{\Delta t}}{2} X_2.$$

– Step $n+1$. Leave all $B_{t_{n+1,2k}} = B_{t_{n,k}}$, for any $k \in \{0,\dots,2^n\}$. Set $\Delta t = T/2^n$ and add, for all $k \in \{1,\dots,2^n\}$ (i.e. for each interval), the value

$$B_{t_{n+1,2k-1}} = (B_{t_{n+1,2k-2}} + B_{t_{n+1,2k}})/2 + \frac{\sqrt{\Delta t}}{2} X_j,$$

where X_j is the next value in the sequence ahead.

The advantage of Brownian bridge sampling is that it is not necessary to give the distance between time points in advance. Further, by this method additional process values can be added to an already simulated sample path, if necessary. Figure 1.2 shows one simulated sample path for $n = 4, 5, 6$ (left) and $n = 10$ (right).

1.3.3. *Karhunen–Loève approximation*

In the previous algorithms, each simulated sample path is a piecewise linear function. In the theory of stochastic processes, the Karhunen–Loève theorem is a representation of a stochastic process as an infinite linear combination of orthogonal functions, analogous to a Fourier series representation of a function on a bounded interval. In contrast to a Fourier series where the coefficients are fixed numbers and the expansion basis consists of sinusoidal functions, the coefficients in the Karhunen–Loève theorem are random variables and the expansion basis depends on the process.

In the case of a centered stochastic process $(Y_t)_{t\in[a,b]}$, Y_t admits a decomposition

$$Y_t = \sum_{k=1}^{\infty} Z_k e_k(t),$$

where $(Z_k)_{k\geq 1}$ is a sequence of pairwise uncorrelated random variables, and where $(e_k)_{k\geq 1}$ is a sequence of continuous real-valued functions on $[a,b]$ that are pairwise orthogonal in $L^2([a,b])$. Moreover, if the process is Gaussian, then the random variables Z_k are Gaussian and independent.

The Karhunen–Loève expansion of a Brownian motion in $[0,T]$ is given by (see [ASH 90], for instance):

$$B_t = \sum_{k=1}^{\infty} Z_k e_k(t) \quad \text{with} \quad e_k(t) = \frac{2\sqrt{2T}}{(2k+1)\pi} \sin\left(\frac{(2k+1)\pi t}{2T}\right) \qquad [1.1]$$

where $(Z_k)_{k\geq 1}$ is a sequence of i.i.d. standard Gaussian random variables. To approximate a sample path of the Brownian motion, we can truncate the series in equation [1.1] up to the N-th term. Figure 1.3 shows some simulated sample paths for different N.

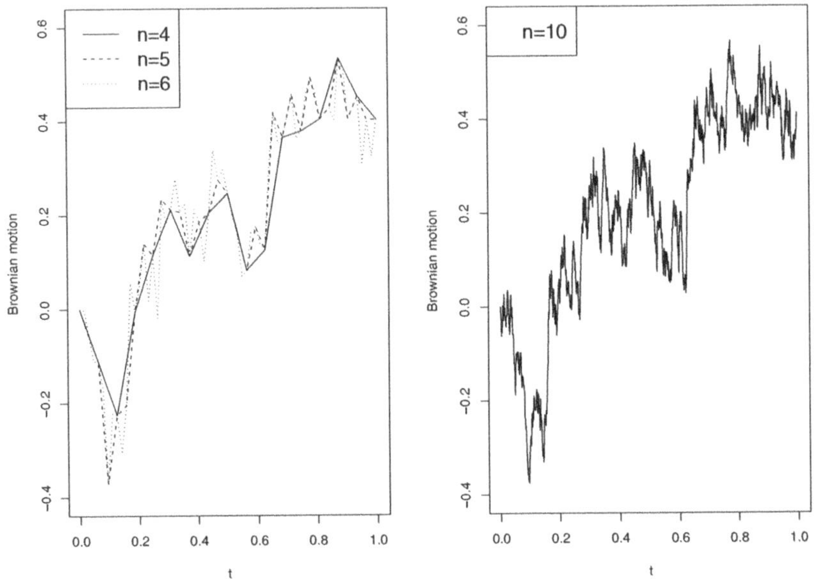

Figure 1.2. *Brownian bridge sampling*

REMARK 1.10.– All three methods of simulation give good results. The usage depends on the concrete situation. The random walk approximation is simple and fast, but the times for process values must be given in advance. In the Brownian bridge sampling, it is possible to add process values to an already simulated sample path, if necessary. By these two methods, the process values are simulated at discrete time points. This is the major difference to the Karhunen–Loève approximation where the sample path is not piecewise linear.

1.4. Wiener process (Brownian motion with drift)

A Wiener process is a Brownian motion with an additional linear drift function and an additional variance parameter:

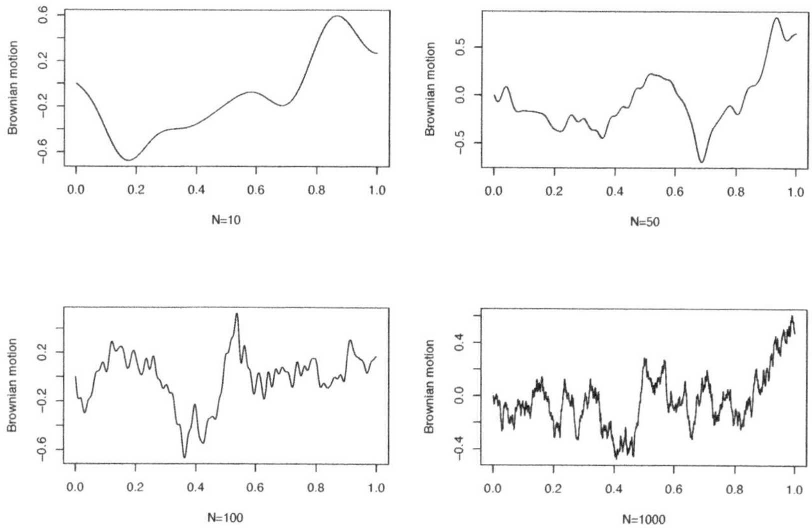

Figure 1.3. *Karhunen–Loève approximation for the Brownian motion*

DEFINITION 1.11.– *The stochastic process* $W = (W_t)_{t \geq 0}$ *is called* a Wiener process *if*

$$\forall t \geq 0, \quad W_t = \mu t + \sigma B_t, \qquad [1.2]$$

where $\mu \in \mathbb{R}$, $\sigma > 0$ *and* $(B_t)_{t \geq 0}$ *is a Brownian motion:* μ *is a drift parameter (that we expect to be positive in our case) and* σ^2 *is a variance parameter (or volatility parameter).*

The Wiener process with drift μ has the same properties as the Brownian motion. In this way, the increments remain independent. The only difference is some change in the parameters of the Gaussian distributed increments: For all $s < t$, the increment $W_t - W_s$ is Gaussian distributed with

$$\mathbb{E}(W_t - W_s) = \mu(t - s) \quad \text{and} \quad \text{Var}(W_t - W_s) = \sigma^2(t - s). \qquad [1.3]$$

It is sometimes useful to make the model more flexible by introducing two additional parameters x_0 and t_0, with $x_0,\ t_0 \in \mathbb{R}$. Then the degradation measure $(X_t)_{t \geq 0}$ can be defined by the following model:

$$\forall t \geq t_0, \quad X_t = x_0 + W_{t-t_0} = x_0 + \mu(t - t_0) + \sigma B_{t-t_0}, \qquad [1.4]$$

where $x_0 \in \mathbb{R}$ is a constant initial degradation, $t_0 \in \mathbb{R}$ is the time when the degradation accumulation begins and $(W_t)_{t\geq 0}$ is a Wiener process.

Model [1.4] differs from [1.2] in the time where the degradation begins to accumulate. It changes the initial time from 0 to t_0 and the initial state from 0 to x_0. The influence of the parameters x_0 and t_0 is illustrated in Figure 1.4. If $t_0 < 0$, then at time $t = 0$ (time when the use of the item begins), the state of the item is random. More specifically, it is Gaussian distributed with mean $x_0 + \mu|t_0|$ and variance $\sigma^2|t_0|$. Assuming h to be a critical level for the deterioration, we will see later on that it is not the boundary h itself, but the difference $h - x_0$ (degradation reserve) that plays an important role. When $t_0 < 0$, the random state at time $t = 0$ provides a random degradation reserve at the same time. An alternate possibility is to directly model h as a random boundary. On the other side, if $t_0 > 0$, the time t_0 can be seen as an "incubation time". It is also possible that we begin the observation of the process after it has been running for a while, and that we are interested in its future development. If there is a short period of nonlinear behavior at the beginning, as is plotted in Figure 1.4(b), and if the observation begins after that time, when the mean function has become linear, then the short nonlinearity period is absorbed by (probably false) estimates of x_0 and t_0, which are not of further interest. This makes the model more flexible.

For any $t \geq t_0$, the p.d.f. of X_t is then

$$\forall x \in \mathbb{R}, \quad f_{X_t}(x) = \varphi\left(\frac{x - x_0 - \mu(t - t_0)}{\sigma(t - t_0)}\right), \qquad [1.5]$$

where φ is the p.d.f. of the standard Gaussian distribution.

1.5. First passage time (time to failure)

We suppose that a failure of a product will occur if the degradation process arrives at a certain critical degradation level, which may be unknown. For a given constant boundary h, the lifetime τ_h of the product is then determined as the first instant at which the degradation process $(X_t)_{t\geq 0}$ exceeds the level h:

$$\tau_h = \inf\{t \geq t_0; X_t \geq h\}. \qquad [1.6]$$

In case $x_0 \geq h$, it is readily seen that $\tau_h = t_0$. We consequently only consider the case where $x_0 < h$ in the following.

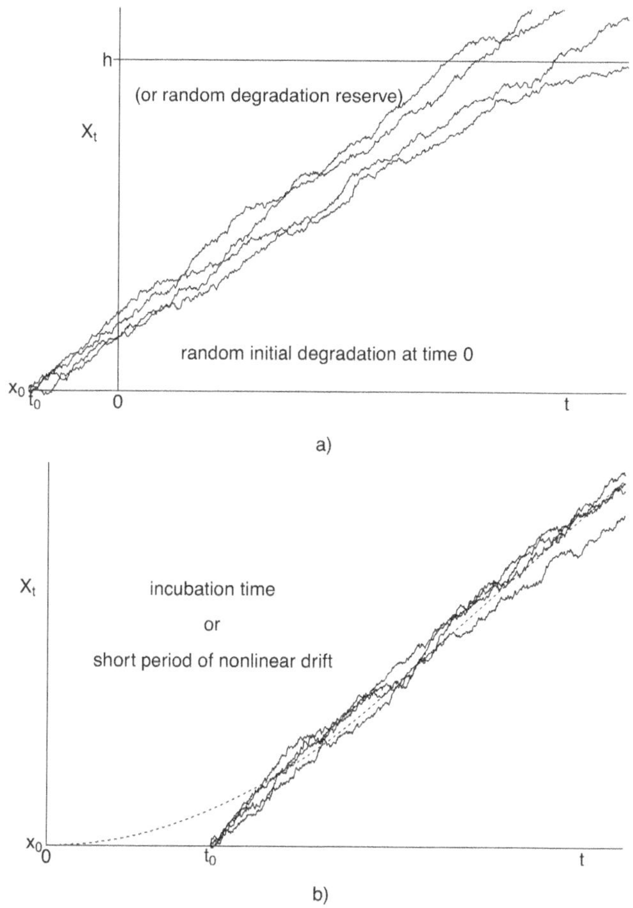

Figure 1.4. *Random initial degradation a) or incubation time b)*

Let $t > t_0$. Noticing that $X_t > x$ with $x > h$ implies that $\tau_h < t$, the formula of total probability implies that, for any $x > h$:

$$\mathbb{P}(X_t > x) = \mathbb{P}(X_t > x; \tau_h < t) = \int_{t_0}^{t} \mathbb{P}(X_t > x | \tau_h = s)\, \mathrm{d}F_{\tau_h}(s).$$

Based on the continuity of $(X_t)_{t\geq 0}$, we also have $X_{\tau_h} = h$. Using furthermore the independent and stationary increments property, we obtain

$$\begin{aligned}\mathbb{P}(X_t > x) &= \int_{t_0}^{t} \mathbb{P}(X_t > x | X_s = h)\,\mathrm{d}F_{\tau_h}(s)\\ &= \int_{t_0}^{t} \mathbb{P}(X_t - X_s > x - h)\,\mathrm{d}F_{\tau_h}(s)\\ &= \int_{t_0}^{t} \mathbb{P}(X_{t-s} > x - h)\,\mathrm{d}F_{\tau_h}(s).\end{aligned} \quad [1.7]$$

Equation [1.7] says that the Wiener process is above state $x > h$ at time t if it takes s units of time to reach level h for the first time and next, the increment during a time interval with length $t - s$ is above $x - h$ (see Figure 1.5).

Taking the derivative of [1.7] with respect to x, we get an integral equation for the p.d.f. of the first passage time:

$$f_{X_t}(x) = \int_{t_0}^{t} f_{\tau_h}(s) f_{X_{t-s}}(x-h)\,\mathrm{d}s, \quad [1.8]$$

where $f_{X_{t-s}}$ is the p.d.f. of the increment of the Wiener process during time $t - s$ and f_{τ_h} the p.d.f. of τ_h.

Let us first consider the solution of the integral equation [1.7] (or equivalently of [1.8]) for the special case $\mu = 0$, $x_0 = 0$ and $t_0 = 0$ (which corresponds to a Brownian motion).

PROPOSITION 1.12.– For $\mu = 0$, $x_0 = 0$ and $t_0 = 0$,

$$\forall t \geq 0, \quad F_{\tau_h}(t) = 2\left(1 - \Phi\left(\frac{h}{\sigma\sqrt{t}}\right)\right)$$

is the unique solution of equation [1.7]. The p.d.f. is then given by:

$$\forall t \geq 0, \quad f_{\tau_h}(t) = \frac{h}{\sqrt{2\pi\sigma^2 t^3}} \exp\left(-\frac{h^2}{2\sigma^2 t}\right). \quad [1.9]$$

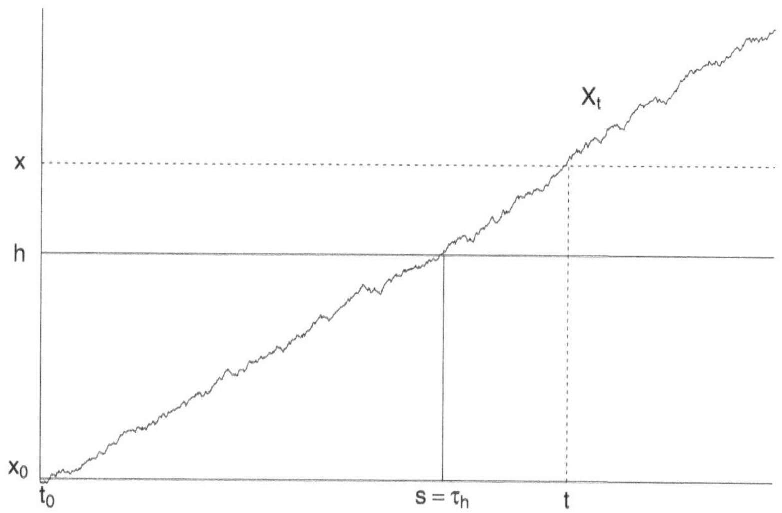

Figure 1.5. *A sample path of the Wiener process*

PROOF.– For these parameters, we have $X_t = \sigma B_t$ and, since $\frac{B_t}{\sqrt{t}}$ is standard Gaussian distributed,

$$\mathbb{P}(X_t > x) = \mathbb{P}(\sigma B_t > x) = \mathbb{P}\left(\frac{B_t}{\sqrt{t}} > \frac{x}{\sigma\sqrt{t}}\right) = 1 - \Phi\left(\frac{x}{\sigma\sqrt{t}}\right),$$

where Φ is the cumulative distribution function of the standard Gaussian distribution. Furthermore, for $x = h$, we have

$$\mathbb{P}(X_{t-s} > x - h) = F_{X_{t-s}}(0) = \frac{1}{2},$$

because X_{t-s} is Gaussian distributed with zero expectation. Now, we get, from equation [1.7],

$$\mathbb{P}(X_t > h) = \int_0^t \frac{1}{2}\,\mathrm{d}F_{\tau_h}(s) = \frac{1}{2}F_{\tau_h}(t)$$

and the c.d.f. of the first passage time is

$$F_{\tau_h}(t) = 2\,\mathbb{P}(X_t > h) = 2\left(1 - \Phi\left(\frac{h}{\sigma\sqrt{t}}\right)\right).$$

Finally, taking the derivative, the p.d.f. of τ_h is given by:

$$\forall t \geq 0, f_{\tau_h}(t) = \frac{\mathrm{d}F_{\tau_h}(t)}{\mathrm{d}t} = \frac{h}{\sigma\sqrt{t^3}}\Phi'\left(\frac{h}{\sigma\sqrt{t}}\right) = \frac{h}{\sqrt{2\pi\sigma^2 t^3}}\exp\left(-\frac{h^2}{2\sigma^2 t}\right).$$

□

From equation [1.9], we can observe that the first passage time of a Brownian motion is inverse Gaussian distributed.

DEFINITION 1.13.– *The inverse Gaussian distribution (also known as the Wald distribution) is a two-parameter family of continuous probability distributions with support on* $(0, \infty)$. *Its p.d.f. is given by:*

$$\forall t \geq 0, \quad f(t) = \sqrt{\frac{\beta}{2\pi t^3}} \exp\left(\frac{-\beta(t-\mu_1)^2}{2\mu_1^2 t}\right), \qquad [1.10]$$

where $\mu_1 > 0$ *is the mean and* $\beta > 0$ *is the shape parameter.*

It is much more complicated to develop the p.d.f. of the first passage time in the case of a general Wiener process with drift. However, the previous result on the Brownian motion gives us an idea of the general solution and we now show that the first passage time for a fixed level $h > x_0$ of a Wiener process with positive drift is still inverse Gaussian distributed.

THEOREM 1.14.– Remembering that the distribution of X_t is given by equation [1.5], then

$$\forall t \geq t_0, \; f_{\tau_h}(t) = \frac{h - x_0}{\sqrt{2\pi\sigma^2(t-t_0)^3}} \exp\left(-\frac{(h - x_0 - \mu(t-t_0))^2}{2\sigma^2(t-t_0)}\right). \qquad [1.11]$$

PROOF.– First, we prove that [1.11] fulfills equation [1.8]. For this, we have to show that

$$\begin{aligned} &\frac{1}{\sqrt{2\pi\sigma^2(t-t_0)}} \exp\left(-\frac{(x - x_0 - \mu(t-t_0))^2}{2\sigma^2(t-t_0)}\right) \\ &= \int_{t_0}^{t} \frac{h - x_0}{\sqrt{2\pi\sigma^2(s-t_0)^3}} \exp\left(-\frac{(h - x_0 - \mu(s-t_0))^2}{2\sigma^2(s-t_0)}\right) \\ &\quad \times \frac{1}{\sqrt{2\pi\sigma^2(t-s)}} \exp\left(-\frac{(x - h - \mu(t-s))^2}{2\sigma^2(t-s)}\right) \mathrm{d}s. \end{aligned}$$

The right-hand side of this equation can be written as:

$$\frac{1}{\sqrt{2\pi\sigma^2(t-t_0)}}\exp\left(-\frac{(x-x_0-\mu(t-t_0))^2}{2\sigma^2(t-t_0)}\right)\times I,$$

where

$$I=\int_{t_0}^{t}\frac{1}{\sqrt{2\pi\sigma^2}}\frac{(h-x_0)(t-t_0)^{1/2}}{(t-s)^{1/2}(s-t_0)^{3/2}}\exp\left(-\frac{1}{2\sigma^2}\cdot A(s)\right)\,\mathrm{d}s,$$

and

$$\begin{aligned}A(s)&=\frac{(h-x_0)^2}{(s-t_0)}+\frac{(x-h)^2}{(t-s)}-\frac{(x-x_0)^2}{(t-t_0)}\\&=\frac{((x-h)(s-t_0)-(h-x_0)(t-s))^2}{(s-t_0)(t-s)(t-t_0)}\\&=\frac{((x-h)\frac{(s-t_0)}{(t-s)}-(h-x_0))^2}{\frac{(s-t_0)}{(t-s)}(t-t_0)}.\end{aligned}$$

Applying the substitution $u=\frac{s-t_0}{t-s}$ for $t_0<s<t$, that is $s=\frac{t_0+ut}{1+u}$ with $0<u<\infty$, we obtain:

$$I=\int_0^{\infty}\frac{1}{\sqrt{2\pi u^3}}\frac{(h-x_0)}{\sigma(t-t_0)^{1/2}}\exp\left(-\frac{(x-h)^2(u-\frac{(h-x_0)}{(x-h)})^2}{2\sigma^2(t-t_0)u}\right)\mathrm{d}u.$$

It is easy to see that this is an integral over the complete support of an inverse Gaussian distribution with $\beta=(h-x_0)^2/(\sigma^2(t-t_0))$ and $\mu_1=(h-x_0)/(x-h)$. Therefore, $I=1$ and [1.11] fulfills equation [1.8].

We now prove that the solution of [1.8] is unique. This can be shown by using the Laplace transforms of $t\mapsto f_{\mathcal{T}_h}(t)$ and $t\mapsto f_{X_t}(x)$:

$$\mathcal{L}_{\mathcal{T}_h}(s;x_0,h)=\int_{t_0}^{\infty}e^{-st}f_{\mathcal{T}_h}(t)\,\mathrm{d}t,\quad \mathcal{L}_X(x;x_0,s)=\int_{t_0}^{\infty}e^{-st}f_{X_t}(x)\,\mathrm{d}t.$$

Note that $\mathcal{L}_{\mathcal{T}_h}(s)$ is a moment generating function whereas $\mathcal{L}_{X_t}(s)$ is not, the integration being taken with respect to t. The integral in the right-side of equation [1.8] can be seen as a convolution product. As the Laplace transform of a convolution product is the product of the Laplace transforms, we derive that

$$\mathcal{L}_{\tau_h}(s; x_0, h) = \frac{\mathcal{L}_X(x; x_0, s)}{\mathcal{L}_X(x; h, s)}.$$

There is hence one single possible Laplace transform of f_{τ_h} for f_{τ_h} to be a solution of [1.8]. The uniqueness of the solution of [1.8] hence follows from the one-to-one property of Laplace transform. See [CHH 89] for more details and a proof in a more general setting (positive, negative, and zero drift). □

We now make some remarks and give some properties of the p.d.f. given in equation [1.11]. More about the inverse Gaussian distribution can be found, for example, in [CHH 89] or in [JØR 82].

1) Setting

$$\begin{cases} \beta &= \frac{(h-x_0)^2}{\sigma^2} \\ \mu_1 &= \frac{h-x_0}{\mu} \end{cases} \Longleftrightarrow \begin{cases} \sigma^2 &= \frac{(h-x_0)^2}{\beta} \\ \mu &= \frac{h-x_0}{\mu_1} \end{cases}, \qquad [1.12]$$

we can see that equations [1.11] and [1.10] are equivalent for $t_0 = 0$ so that, indeed, τ_h is inverse Gaussian distributed with parameter (β, μ_1) in the sense of definition 1.13. $t_0 \in \mathbb{R}$ is an additional location parameter and we get the three-parameter inverse Gaussian distribution (see [PAD 79] for instance).

2) The c.d.f. of τ_h is given by, for any $t \geq t_0$,

$$F_{\tau_h}(t) = \mathbb{P}(\tau_h \leq t) = \Phi\left(\frac{\mu(t-t_0) - h + x_0}{\sigma\sqrt{t-t_0}}\right) + \exp\left(\frac{2\mu(h-x_0)}{\sigma^2}\right)\Phi\left(-\frac{\mu(t-t_0) + h - x_0}{\sigma\sqrt{t-t_0}}\right).$$

This can be checked by taking its derivative that leads to equation [1.11].

3) The characteristic function $C_{\tilde{\tau}_h}(z) = \mathbb{E}(e^{iz\tilde{\tau}_h})$, where $\tilde{\tau}_h = \tau_h - t_0$, is given by

$$C_{\tilde{\tau}_h}(z) = \exp\left(\frac{\beta}{\mu_1}\left(1 - \left(1 - \frac{2i\mu_1^2 z}{\beta}\right)^{1/2}\right)\right),$$

where (β, μ_1) is provided by [1.12]. Taking the k-th derivative of $C_{\tilde{\tau}_h}(z)$ and evaluating it at $z = 0$, we get (see, e.g. [CHH 89]):

$$\mathbb{E}[\tilde{\tau}_h^k] = \mu_1^k \sum_{u=0}^{k-1} \frac{(k-1+u)!}{u!((k-1-u)!} \left(\frac{\mu_1}{2\beta}\right)^u . \qquad [1.13]$$

It follows that the expectation and the variance of τ_h are given by:

$$\mathbb{E}(\tau_h) = t_0 + \mu_1 = t_0 + \frac{h - x_0}{\mu} \text{ and } \mathrm{Var}(\tau_h) = \frac{\mu_1^3}{\beta} = \frac{(h - x_0)\sigma^2}{\mu^3}.$$

Note that for a Brownian motion without drift, that is $\mu = 0$, the expectation of τ_h does not exist.

4) The failure rate $r(t) = f(t)/(1 - F(t))$ is rather complicated, but it is not difficult to compute it for given parameters. The behavior of the failure rate was analyzed in detail in [CHH 89]. It is increasing for $t < t^*$ and decreasing for $t > t^*$, where t^* is the solution of

$$r(t) = \frac{\beta}{2\mu_1^2} + \frac{3}{2t} - \frac{\beta}{2t^2}.$$

Furthermore, it can be shown that

$$\lim_{t\to\infty} r(t) = \frac{\beta}{2\mu_1^2}.$$

At first look, it might seem surprising that the inverse Gaussian distribution does not have an increasing failure rate. The reason becomes clearer if we think in terms of the corresponding parameters of the Wiener process. In [CHH 89, p. 153], some failure rates are plotted for $\mu_1 = 1$ and different β. It can be seen that the mode t^* of $r(t)$ is small for β small. Since $\beta = (h - x_0)^2/\sigma^2$, the mode of $r(t)$ is hence small either if the distance between the initial value x_0 and the boundary h is small, or if the variance σ^2 of the process is very large. In both cases, the stochastic process either reaches the boundary level after a short time or it first goes down and needs some time to come back. The decreasing property of the failure rate after a short time can hence be justified in terms of the physical behavior: excluding the items that are reaching the boundary after a short time, the failure rate of the surviving items will be relatively small. In Figure 1.6, some failure rates together with their c.d.f. for $h - x_0 = 5$, $\mu = 1$ and different σ are plotted. It can be seen that for small and moderate variances σ^2, the failure rate is increasing, it is not monotonous any more when σ^2 is large.

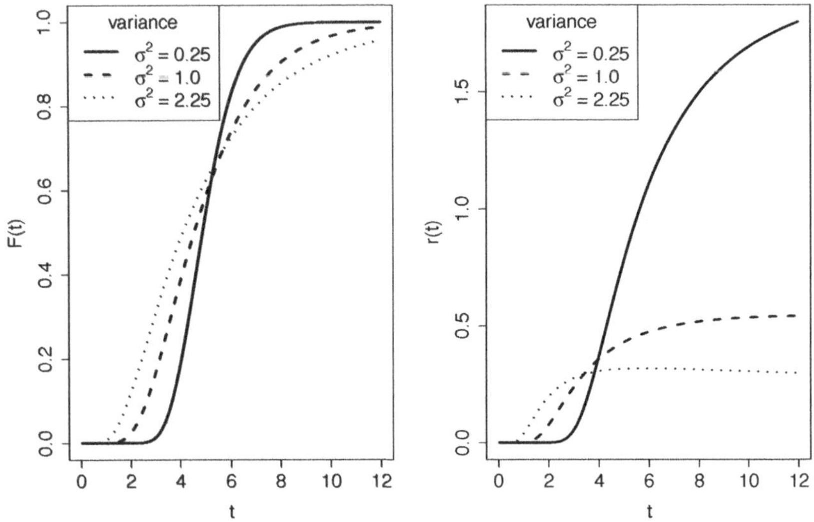

Figure 1.6. *C.d.f. and failure rate of some times-to-failure distributions*

5) Let $\tilde{\tau}_h = \tau_h - t_0$. In [CHH 89], it was shown that for the mean residual life $\mathbb{E}(\tilde{\tau}_h - T|\tilde{\tau}_h > T)$ of an item with age T, the following properties hold:

$$\lim_{T\to 0} \frac{\mathbb{E}(\tilde{\tau}_h - T|\tilde{\tau}_h > T)}{\mathbb{E}(\tilde{\tau}_h)} = 1 \text{ and}$$

$$\lim_{T\to\infty} \frac{\mathbb{E}(\tilde{\tau}_h - T|\tilde{\tau}_h > T)}{\mathbb{E}(\tilde{\tau}_h)} = \frac{2\mu_1}{\beta} = \frac{2\sigma^2}{\mu(h - x_0)}.$$

When σ^2 is large, the mean residual lifetime can hence exceed the mean initial lifetime of the system (obtained when $T \to 0$). This apparently surprising property can be explained in a similar way as for the failure rate.

1.6. Statistical inference

It was shown that for $x_0 < h$, the lifetime τ_h follows an inverse Gaussian distribution with p.d.f. given in equation [1.11]. The parameters of this distribution are expressed with respect to those of the degradation process using [1.12] so that estimating the parameters of the inverse Gaussian distribution is mostly equivalent to estimating the parameters of the Wiener process (details further). The degradation level h may also be an unknown parameter, which may hence have to be estimated too. To estimate the parameters of this model, several observations schemes are possible:

1) Only increments of the degradation process are observed. In this case, the likelihood function of the observation is a product of Gaussian densities. It is possible to estimate the parameters of the degradation process μ, σ^2, x_0 and t_0, but it is impossible to estimate the degradation level h. Note that, neither the same delay between successive observation times, nor identical observation times, are required. Parameter estimations can be obtained even if the observation times are all distinct.

2) Only failure times are observed. In this case, we get the classical likelihood function from a sample of inverse Gaussian distributed random variables. From this likelihood function, we can estimate the parameters μ, σ^2 and t_0. The degradation reserve $h - x_0$ must be given in advance.

3) The sample includes both process increments and failure times. Assume first that the sample consists of N failure times and of $\sum_{i=1}^{n} m_i$ increments corresponding to different degradation processes from those on which the failure times are observed. Then the likelihood function is the product of the likelihood functions mentioned above. In the case where failure times and degradation measures correspond to the same process, the likelihood function is more complicated to be computed. In this case, the p.d.f. of an observed process increment is a conditional density under the assumption that the process does not cross the degradation level in the current time interval. If the sample includes both process increments and failure times, then it is possible to estimate all parameters of the model.

These models and their parameter estimation are described in [KAH 04], [KAH 98] and in [KAH 10a]. A similar model and its application in medicine is described in [DOK 96]. Several generalizations of this simple model were given. It is possible to include measurement errors [WHI 95], or to transform the time scale [WHI 97]. It is also possible to consider bivariate models [WHI 98].

Here, we consider parameter estimation for the three different observation levels. In section 1.6.1, only degradation data are considered. We will find estimators for the parameters of a Wiener process with drift together with the two parameters describing the starting time and initial state of the process. Furthermore, confidence regions and improved confidence regions are constructed. In the short section 1.6.2, we consider the classical sample of lifetimes. Finally, in section 1.6.3, we consider the case where, for each stochastic process, both degradation measures and a failure time are observable, and we estimate simultaneously the parameters μ, σ^2, t_0, x_0 and h. By contrast with the previous observations schemes, here, for each sample path of the degradation process, either a failure time is observed or degradation data at fixed observation times under the condition that the process has not yet exceeded the degradation level h. Hence, these conditional process increments and the failure time τ_h are dependent random variables. To compute the likelihood function, we have to find the conditional distribution of the process under the condition that the level h has not yet been exceeded, together with the joint distribution of conditional process increments and the lifetime τ_h.

1.6.1. *Statistical inference for degradation data*

In this section, we assume that the process is not stopped by a failure (which is equivalent to assuming that h tends to infinity). The observation of the underlying degradation process at fixed times gives an alternative approach to parameter estimation from lifetime data, which is particular useful for products with high reliability, since failure times have not to be observed. If we ignore the boundary level h, the parameters μ, σ^2, t_0 and x_0 may be estimated by use of data of the degradation process before a failure occurs. For this case, an example of a sample is given in Figure 1.7.

The sampling scheme is the following one: n i.i.d. copies of the degradation process $(X_t^{(1)})_{t \geq t_0}, \ldots, (X_t^{(n)})_{t \geq t_0}$ are observed. In addition, there are m_i observations of the i-th Wiener process at times $t_{i,1}, \ldots, t_{i,m_i}$. We now introduce several notations:

– $Y_{i,j} = X_{t_{i,j}}^{(i)} - X_{t_{i,j-1}}^{(i)}$: increments of the process;

– for $j \in \{2, \ldots, m_i\}$, $s_{i,j} = t_{i,j} - t_{i,j-1}$: delay between observation times;

– $X_{i,1} = X_{t_{i,1}}^{(i)}$: first observed degradation measure;

– $\bar{X}_{\cdot 1} = \frac{1}{n} \sum_{i=1}^{n} X_{t_{i,1}}^{(i)}$: average value of the first observed degradation measures;

– $\bar{X}_{\cdot m} = \frac{1}{n} \sum_{i=1}^{n} X_{t_{m_i}}^{(i)}$: average value of the last observed degradation measures;

– $\bar{t}_{\cdot 1} = \frac{1}{n} \sum_{i=1}^{n} t_{i,1}$: average value of the first observation times;

– $\bar{t}_{\cdot m} = \frac{1}{n}\sum_{i=1}^{n} t_{i,m_i}$: average value of the last observation times;

– $\bar{m} = \frac{1}{n}\sum_{i=1}^{n} m_i$: average value of the number of observations per process.

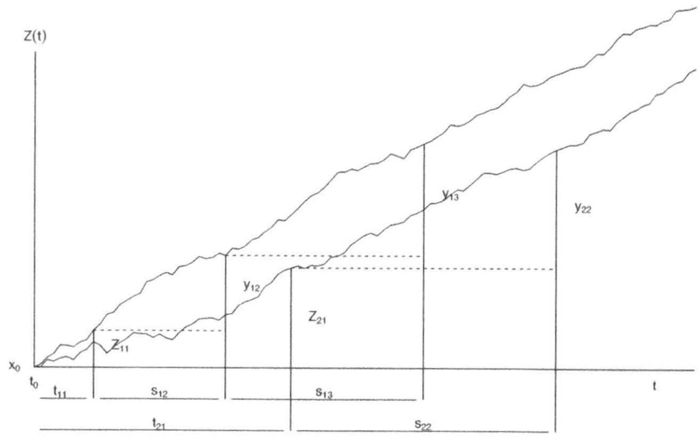

Figure 1.7. *Sample for observations of process increments for the case* $n = 2$, $m_1 = 3$, $m_2 = 2$

Lower case letter notations such as $y_{i,j}$ denote observed values of the corresponding random variables. Further, let $\mathbf{D}_{obs} = \bigcup_{i=1}^{n}\bigcup_{j=1}^{m_i} \{(t_{i,j}, X_{i,j})\}$ be the set of all observed degradation data. The increments are independent and Gaussian distributed:

$$X_{i,1} \sim \mathcal{N}(x_0 + \mu(t_{i,1} - t_0), \sigma^2(t_{i,1} - t_0)), \quad \forall i \in \{1, \dots, n\}$$
$$Y_{i,j} \sim \mathcal{N}(\mu s_{i,j}, \sigma^2 s_{i,j}), \quad \forall j \in \{1, \dots, m_i\}, \forall i \in \{1, \dots, n\}$$

and the resulting likelihood function has the form:

$$L(x_0, t_0, \mu, \sigma^2 | \mathbf{D}_{obs}) = \prod_{i=1}^{n} \frac{1}{\sqrt{\sigma^2(t_{i,1} - t_0)}} \varphi\left(\frac{x_{i,1} - x_0 - \mu(t_{i,1} - t_0)}{\sqrt{\sigma^2(t_{i,1} - t_0)}}\right)$$
$$\times \prod_{i=1}^{n}\prod_{j=2}^{m_i} \frac{1}{\sqrt{\sigma^2 s_{i,j}}} \varphi\left(\frac{y_{i,j} - \mu s_{i,j}}{\sqrt{\sigma^2 s_{i,j}}}\right), \qquad [1.14]$$

where φ is p.d.f. of the standard normal distribution. Here, it has to be supposed that all the first observation times $t_{1,1}, \dots, t_{n,1}$ are greater than t_0.

In this case, which is formally obtained on setting $h = \infty$, we do not make use of information on failure times nor of the fact that the degradation process does not exceed the level h between two successive observation times. The maximum likelihood equations are as follows:

$$\begin{cases} \widehat{\mu} = (\bar{x}_{\cdot m} - \widehat{x}_0)/(\bar{t}_{\cdot m} - \widehat{t}_0) \\ \widehat{\sigma}^2 = \dfrac{1}{\bar{m}n}\left(\sum_{i=1}^{n}\left(\dfrac{(x_{i,1} - \widehat{x}_0)^2}{t_{i,1} - \widehat{t}_0} + \sum_{j=2}^{m_i}\dfrac{y_{i,j}^2}{s_{i,j}}\right) - \widehat{\mu}n(\bar{x}_{\cdot m} - \widehat{x}_0)\right) \\ \widehat{x}_0 = \left(\dfrac{1}{n}\sum_{i=1}^{n}\dfrac{x_{i,1}}{t_{i,1} - \widehat{t}_0} - \dfrac{\bar{x}_{\cdot m}}{\bar{t}_{\cdot m} - \widehat{t}_0}\right) \Bigg/ \left(\dfrac{1}{n}\sum_{i=1}^{n}\dfrac{1}{t_{i,1} - \widehat{t}_0} - \dfrac{1}{\bar{t}_{\cdot m} - \widehat{t}_0}\right) \\ \widehat{\sigma}^2 \sum_{i=1}^{n}\dfrac{1}{t_{i,1} - \widehat{t}_0} = \sum_{i=1}^{n}\dfrac{(x_{i,1} - \widehat{x}_0)^2}{(t_{i,1} - \widehat{t}_0)^2} - \widehat{\mu}^2 n. \end{cases}$$

The last two equations, after replacing $\widehat{\mu}$ and $\widehat{\sigma}^2$ in the last equation, can be solved numerically with respect to $\widehat{x}_0$ and $\widehat{t}_0$. If these estimates are found, we get estimates $\widehat{\mu}$ and $\widehat{\sigma}^2$ for the two process parameters. An essential simplification arises if the first observation times are identical for all degradation process: $t_{i,1} = t_1,\ i \in \{1, \ldots, n\}$. Then, an explicit solution of the likelihood equations can be found:

$$\begin{cases} \widehat{\mu} = (\bar{x}_{\cdot m} - \bar{x}_{\cdot 1})/(\bar{t}_{\cdot m} - t_1) \\ \widehat{\sigma}^2 = \dfrac{1}{\bar{m}(n-1)}\left(\sum_{i=1}^{n}\sum_{j=2}^{m_i}\dfrac{y_{i,j}^2}{s_{i,j}} - n\dfrac{(\bar{x}_{\cdot m} - \bar{x}_{\cdot 1})^2}{\bar{t}_{\cdot m} - t_1}\right) \\ \widehat{x}_0 = \left(\bar{x}_{\cdot 1}(\bar{t}_{\cdot m} - \widehat{t}_0) - \bar{x}_{\cdot m}(t_1 - \widehat{t}_0)\right)/(\bar{t}_{\cdot m} - t_1) \\ \widehat{t}_0 = t_1 - \dfrac{1}{\widehat{\sigma}^2}\left(\dfrac{1}{n}\sum_{i=1}^{n} x_{i,1}^2 - \bar{x}_{\cdot 1}^2\right). \end{cases} \qquad [1.15]$$

In the following, we will only consider the simpler case of equal first observation times. Then, the likelihood function has some "nice" properties. First, it can be proved, see [KAH 04], that the estimators given in equation [1.15] are asymptotically Gaussian distributed if:

– either, for estimating all parameters, the number of observed processes n tends to infinity;

– or, for estimating μ and σ^2, we observe only one process, but $\bar{m}$ and $(\bar{t}_{\cdot m} - t_1)$ tend to infinity.

For the situation where the process is already running for a while, it may appear that only the parameters μ and σ^2 are of interest, while x_0 and t_0 (which describe the beginning of the degradation) are nuisance parameters. If the first observation times are identical for all degradation processes, then new nuisance parameters ν_1 and ν_2 can be found, which are independent of the parameters of interest μ and σ^2. This can be seen from the structure of the likelihood function given by equation [1.14]: x_0 and t_0 are present only in the first part. The reparameterization

$$\begin{cases} \nu_1 = x_0 + \mu(t_1 - t_0) \\ \nu_2 = \sigma^2(t_1 - t_0) \end{cases}$$

leads to the likelihood function

$$L(\mu, \sigma^2, \nu_1, \nu_2 | \mathbf{D}_{obs}) = \prod_{i=1}^{n} \frac{1}{\sqrt{\nu_2}} \varphi\left(\frac{x_{i,1} - \nu_1}{\sqrt{\nu_2}}\right) \prod_{i=1}^{n} \prod_{j=2}^{m_i} \times \frac{1}{\sqrt{\sigma^2 s_{i,j}}} \varphi\left(\frac{y_{i,j} - \mu s_{i,j}}{\sqrt{\sigma^2 s_{i,j}}}\right). \qquad [1.16]$$

Note that the new parameters describe the expectation and the variance of the degradation process at the first observation time t_1. The likelihood function now consists of two independent parts: one part includes the nuisance parameters, ν_1 and ν_2, while the second part includes the parameters of interest, μ and σ^2. If we are interested in the process parameters, only the last part of the likelihood function must be taken into consideration. The information matrix of the parameter $\theta = (\mu, \sigma^2, x_0, t_0)^T$ can be found to be

$$I_\theta = n \begin{bmatrix} \frac{\bar{t}_{\cdot m} - t_1}{\sigma^2} & 0 & 0 & 0 \\ 0 & \frac{\bar{m} - 1}{2\sigma^4} & 0 & 0 \\ 0 & 0 & \frac{1}{\nu_2} & 0 \\ 0 & 0 & & \frac{1}{2\nu_2^2} \end{bmatrix}. \qquad [1.17]$$

All parameter estimators are asymptotically independent. The fact that the maximum likelihood estimator is asymptotically normally distributed can be used to construct confidence intervals for each parameter or simultaneous confidence regions for some of them. There are two possibilities for constructing asymptotic confidence regions. The first one is based on the Wald test statistic. An isoline of the k-dimensional Gaussian distribution, where k is the number of parameters, can be determined such that inside of this line there is $(1 - \alpha)$ of the probability mass of the distribution. The equation of this level line is given by the quadratic form [KAH 04]:

$$(\widehat{\theta} - \theta)^T I_\theta(\widehat{\theta} - \theta) < a^2, \qquad [1.18]$$

where $\widehat{\theta} = (\widehat{\mu}, \widehat{\sigma}^2, \widehat{x}_0, \widehat{t}_0)^T$ and where a^2 is the quantile of the χ^2-distribution with $k = 4$ degrees of freedom. The second possibility is based on the log-likelihood ratio statistic. It is well-known (see, e.g. [BAR 86] or [COR 87]) that under the same conditions of regularity as for asymptotic normality, the log-likelihood ratio

$$w = 2\left[\log(L(\widehat{\theta}|\mathbf{D}_{obs}) - \log(L(\theta_0|\mathbf{D}_{obs})\right]$$

converges in distribution to a χ^2-distribution with k degrees of freedom, where k is the number of parameters of interest, $\widehat{\theta}$ the maximum likelihood estimator of all parameters and θ_0 is the maximum likelihood estimator of the nuisance parameters, which depend on the parameters of interest. The advantage of the likelihood ratio statistic is that all nuisance parameters are excluded while they are present in the asymptotic variance of MLE. The rate of convergence of the log-likelihood ratio statistic to the χ^2-distribution is $O(n^{-1})$, where n is the sample size, and it may be improved to $O(n^{-3/2})$ by the Bartlett adjustment R if the fourth derivatives of the log-likelihood and their expectations exist. The log-likelihood ratio with Bartlett adjustment has the following form:

$$w' = w/(1 + R/p),$$

where p is the number of parameters of interest and R is the Bartlett adjustment that depends on the expectations of derivatives of the log-likelihood function and of the derivatives of these expectations. This adjustment is quite difficult to calculate in general. It becomes a little simpler if the estimators are asymptotically independent, because then

$\mathbb{E}(\frac{\partial \log L(\theta|\mathbf{D}_{obs})}{\partial\theta_i\partial\theta_j}) = 0$ for $i \neq j$. In [KAH 04], the Bartlett adjustment was found to be $R = 11/(6n(\bar{m}-1))$.

As a particular case, we here discuss a little the special case where $t_0 = 0 = x_0$. For any $i \in \{1,\ldots,n\}$ and any $j \in \{1,\ldots,m_i\}$, let $Y_{i,j} = X^{(i)}_{t_{i,j}} - X^{(i)}_{t_{i,j-1}}$ (with the convention that $t_{i,0} = 0$): $Y_{i,j}$ is normally distributed with mean $\mu s_{i,j}$ and variance $\sigma^2 s_{i,j}$, where $s_{i,j} = t_{i,j} - t_{i,j-1}$. From the independence of $Y_{i,j}$'s, it is straightforward to obtain the likelihood function and it leads to the following estimators:

$$\widehat{\mu} = \frac{\sum_{i=1}^{n}\sum_{j=1}^{m_i} y_{i,j}}{\sum_{i=1}^{n}\sum_{j=1}^{m_i} s_{i,j}} = \frac{\sum_{i=1}^{n} x^{(i)}_{t_{i,m_i}}}{\sum_{i=1}^{n} t_{i,m_i}} \quad \text{and}$$

$$\widehat{\sigma}^2 = \frac{1}{n\bar{m}}\sum_{i=1}^{n}\sum_{j=1}^{m_i}\frac{(y_{i,j}-\widehat{\mu}s_{i,j})^2}{s_{i,j}}. \qquad [1.19]$$

Under the same condition as in the general case, these estimators are also asymptotically normal.

EXAMPLE 1.15.– To compare all possible confidence regions, we consider joint confidence estimators for the process parameters of interest μ and σ^2 for equal first observation times.

Based on the Wald statistic, the confidence region is an isoline of the two-dimensional Gaussian distribution:

$$\frac{(\widehat{\mu}-\mu)^2}{\sigma^2}(\bar{t}_{\cdot m}-t_1) + \frac{(\widehat{\sigma}^2-\sigma^2)^2}{2\sigma^4}(\bar{m}-1) \leq \frac{a^2}{n}, \qquad [1.20]$$

where $a^2 = -2\log\alpha$ is the $(1-\alpha)$-quantile of the χ^2-distribution with two degrees of freedom. The asymptotic variances of the estimators contain the unknown parameter σ^2 and there are two possibilities for dealing with this issue. The first and most usual possibility is to replace the variance by its consistent estimate and equation [1.20] becomes:

$$\frac{(\widehat{\mu}-\mu)^2}{\widehat{\sigma}^2}(\bar{t}_{\cdot m}-t_1) + \frac{(\widehat{\sigma}^2-\sigma^2)^2}{2\widehat{\sigma}^4}(\bar{m}-1) \leq \frac{a^2}{n}. \qquad [1.21]$$

The joint confidence region is an ellipse. A second possibility is to leave the unknown parameter within the asymptotic variance. Then, equation [1.20] is not an ellipse any more, and for small sample sizes the two regions are quite different.

The log-likelihood ratio is given by:

$$\frac{(\widehat{\mu}-\mu)^2}{\sigma^2}(\bar{t}_{\cdot m}-t_1)+\left(\frac{\widehat{\sigma}^2}{\sigma^2}-1-\log\left(\frac{\widehat{\sigma}^2}{\sigma^2}\right)\right)\geq -\frac{2\log(\alpha)}{n}. \quad [1.22]$$

For the Bartlett correction, the right-hand side of equation [1.22] gets the factor $1+11/(12n(\bar{m}-1))$, the confidence region becomes a little larger for small $n(\bar{m}-1)$. The correction factor tends to 1 if the sample size $n(\bar{m}-1)$ tends to infinity.

In Figure 1.8, the confidence regions for a relatively large sample size of 60 observations are plotted. We have simulated $n = 10$ trajectories of the process with $t_0 = x_0 = 0$, $\mu = 1$, $\sigma^2 = 0.5$, $s_{i,j} = 1$ and $\bar{m} = 6$. The Wald statistic with estimated variance leads to an ellipse, it is the smallest area. The difference between likelihood ratio statistic with and without correction factor is very small for this sample size. The star presents the MLE and "+" is the true parameter.

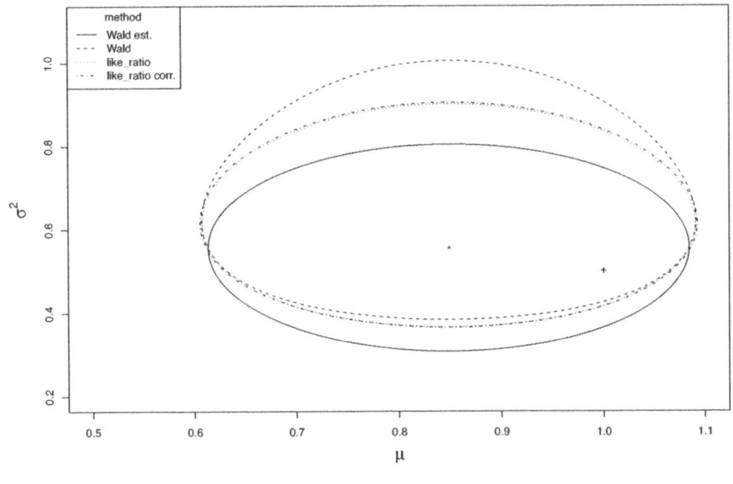

Figure 1.8. *Confidence regions for μ and σ^2 (60 observations)*

In Figure 1.9 are plotted the confidence regions for a relatively small sample size of 25 observations. We have simulated $n = 5$ trajectories of the process with $t_0 = x_0 = 0$, $\mu = 1$, $\sigma^2 = 0.5$, $s_{i,j} = 1$ and $\bar{m} = 5$. Here, we have some difference between likelihood ratio statistic with and without correction factor. The Wald statistic leads to a very large confidence region.

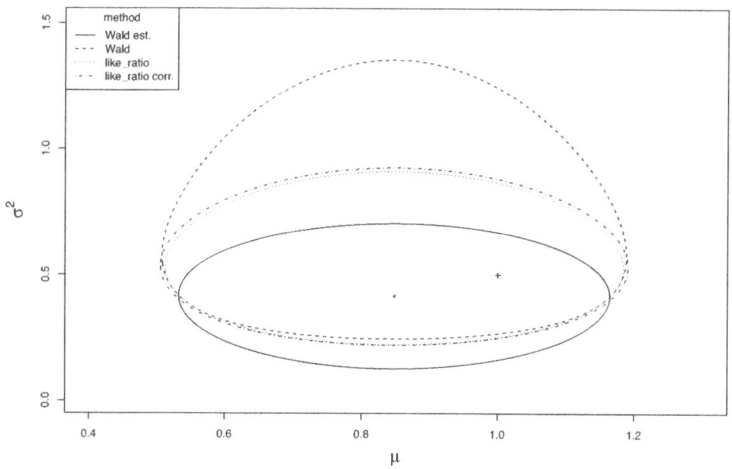

Figure 1.9. *Confidence regions for μ and σ^2 (25 observations)*

To compare the confidence level for finite sample sizes, a small simulation study is performed. We have simulated $n = 5$ trajectories of the process with $t_0 = x_0 = 0$, $\mu = 5$, $\sigma^2 = 0.25$, $s_{i,j} = 1$ and different $\bar{m}$. The number of observations $n\bar{m}$ goes from 15 ($\bar{m} = 3$) to 500 ($\bar{m} = 10$). For each sample size, we have simulated 1,000 samples, estimated the parameters and constructed the four confidence regions given above with confidence level $1-\alpha = 0.95$. We compute the empirical coverage probability. The results are given in Table 1.1.

$\bar{m}$	Equation [1.20]	Equation [1.21]	Equation [1.22]	Equation [1.22] corrected
3	95.40	83.93	93.95	94.82
5	95.19	87.65	94.12	94.68
7	95.27	89.50	94.44	94.74
10	95.51	91.92	95.16	95.43

Table 1.1. *Empirical confidence levels*

It can be seen that the best confidence regions (regarding confidence level), even for small sample sizes, are those based on the likelihood ratio statistic and Wald statistic without replacing the information by its estimate. The replacement, however, leads to a real confidence level smaller than the theoretical. The likelihood ratio statistic has for small sample sizes a smaller real confidence level than it should be, but it becomes better by introducing the Bartlett correction factor. Bearing in mind that the Wald statistic leads to large confidence regions for small sample sizes, the likelihood ratio statistic is the best choice.

REMARK 1.16.– In the next sections, moment method estimators are considered as an alternative to MLE. For estimation based on observations of the Wiener process, the two methods coincide.

1.6.2. *Statistical inference for time to failure data*

When we know that a failure is the result of an underlying degradation process, the time to failure follows an inverse Gaussian distribution with p.d.f. given by equation [1.11]. Since the parameters of this time to failure are expressed with respect to the parameters of the underlying degradation process, it is possible to estimate the degradation parameters from a sample of failure times $\tau_1, \ldots, \tau_n$.

In this case, we get the classical likelihood function from a sample of inverse Gaussian distributed random variables:

$$L(x_0, t_0, \mu, \sigma^2|\mathbf{D}_{obs}) = \prod_{i=1}^{n} \frac{h - x_0}{\sqrt{2\pi\sigma^2(\tau_i - t_0)^3}} \exp\left(-\frac{(h - x_0 - \mu(\tau_i - t_0))^2}{2\sigma^2(\tau_i - t_0)}\right) \mathbf{1}_{\{\tau_i > t_0\}}. \quad [1.23]$$

The estimators for the parameters μ and σ^2 can be calculated explicitly:

$$\widehat{\mu} = \frac{h - x_0}{\frac{1}{n}\sum_{i=1}^{n}(\tau_i - t_0)} \quad \text{and} \quad \widehat{\sigma}^2 = (h - x_0)^2 \frac{1}{n}\sum_{i=1}^{n} \frac{1}{(\tau_i - t_0)} - \widehat{\mu}(h - x_0).$$

Again, it can be shown that these estimators are asymptotically Gaussian distributed as n tends to infinity [KAH 04]. The problem of estimating t_0 is much more complicated, because it is a parameter that defines the support of the lifetime distribution. In [PAD 79], several methods were considered for estimating the parameters of the three-parameter inverse Gaussian distribution by maximum likelihood estimators, moment method estimators and a combination of these two methods.

Note that here we can only estimate the parameters μ, σ^2 and t_0 and that the degradation reserve $h - x_0$ must be given in advance.

1.6.3. *Statistical inference for both degradation and time to failure data*

Let $(X_t)_{t \geq t_0}$ be a degradation process on a probability space $(\Omega, \mathcal{F}, \mathbb{P})$ with values in $(\mathbb{R}, \mathcal{B})$ as defined by equation [1.4], where $\mathcal{B}$ is the Borel σ-algebra on $\mathbb{R}$. Furthermore, let $t_1, \ldots, t_m$ with $m \in \mathbb{N}^*$ be fixed observation times with $t_0 < t_1 < \ldots < t_m < \infty$. To alleviate the notation, we consider only one single degradation process $(X_t)_{t \geq t_0}$ and thus we drop the subscript i. We assume that a failure is observable at any time $t > t_0$ and that we stop observing the degradation process as soon as a failure has occurred. Hence, in each time interval $(t_{j-1}, t_j]$, with $j \in \{1, \ldots, m\}$, we observe either a failure at time $\tau_h \in (t_{j-1}, t_j]$ or we observe the degradation measure $x_j = X_{t_j}$ at time t_j under the condition that the process has not yet exceeded the level h until the time t_j. The observable stopped process is given, for any $t \geq t_0$, by:

$$\widetilde{X}_t = \begin{cases} X_t & \text{if } t_0 \leq t \leq \tau_h, \\ \infty & \text{if } t > \tau_h. \end{cases}$$

Thus, $\widetilde{X}_t$ is a functional of $\{X(s); t_0 \leq s \leq t\}$:

$$\forall t \geq t_0, \quad \widetilde{X}_t = G_t(\tau_h, X_t), \qquad [1.24]$$

where, for any $t, \tau \geq t_0$ and any $x \in \mathbb{R}$,

$$G_t(\tau, x) = \begin{cases} x & \text{if } \tau \geq t \\ \infty & \text{if } \tau < t \end{cases}$$

and $\widetilde{X}_t$ is measurable with respect to the σ-algebra $\mathcal{F}_t = \sigma\{X_s;\ t_0 \leq s \leq t\}$.

A sample of censored observations has the structure:

$$X = F(\tau_h, X_{t_1}, \ldots, X_{t_m}) = (\min(\tau_h, t_m), \widetilde{X}_{t_1}, \ldots, \widetilde{X}_{t_m}) \qquad [1.25]$$

with

$$F(\tau, x_1, \ldots, x_m) = (F_0(\tau), G_{t_1}(\tau, x_1), \ldots, G_{t_m}(\tau, x_m))$$

and

$$F_0(\tau) = \min(\tau, t_m).$$

The statistical model is now given by:

$$(\mathbb{R} \times \overline{\mathbb{R}}^m, \mathcal{B} \otimes \overline{\mathcal{B}}^m, (\mathbb{P}_\theta^F)_{\theta=(\mu,\sigma^2,x_0,t_0,h)\in\Theta\subset\mathbb{R}^5}),$$

where $\overline{\mathbb{R}} = \mathbb{R} \cup \{-\infty, +\infty\}$, $\overline{\mathcal{B}}$ is the σ-algebra of Borel sets in $\overline{\mathbb{R}}$ and $\mathbb{P}_\theta^F$ is the distribution of F when the model parameter is θ.

To compute the likelihood function of X, we have to find the Radon–Nikodym derivative of $\mathbb{P}_\theta^F$ with respect to a dominating measure ν:

$$L(x_0, t_0, \mu, \sigma^2 | \mathbf{D}_{obs}) = \frac{\mathrm{d}\mathbb{P}_\theta^F}{\mathrm{d}\nu}(\tau, \mathbf{x}), \ \tau \in \mathbb{R}, \mathbf{x} = (x_1, \ldots, x_m) \in \overline{\mathbb{R}}^m. \quad [1.26]$$

A crucial step in the derivation of $L(x_0, t_0, \mu, \sigma^2 | \mathbf{D}_{obs})$ is the calculation of the conditional density $z \mapsto f(s, x, t, z, h)$ of the process X_t at time t starting at point $x = X_s$ under the condition that the level h is not exceeded in $[s, t]$. For that purpose, for $s \geq t_0$ and $x \in \mathbb{R}$, let $\mathbb{P}_{s,x}$ stand for the conditional probability $\mathbb{P}_{s,x} = \mathbb{P}(\cdot | X_s = x)$ and let us introduce the transition kernel $(\mathbb{P}(s, x, t, \mathrm{d}y))_{t \geq t_0}$ of the Markov process $(X_t)_{t \geq t_0}$:

$$\begin{aligned}\mathbb{P}(s, x, t, B) &= \mathbb{P}(X_t \in B | X_s = x) \\ &= \frac{1}{\sqrt{2\pi\sigma^2(t-s)}} \int_B \exp\left(-\frac{(y - x - \mu(t-s))^2}{2\sigma^2(t-s)}\right) \mathrm{d}y\end{aligned}$$

for all $t > s \geq t_0$, all $x \in \mathbb{R}$ and all $B \in \mathcal{B}$.

$\mathbb{P}_{s,x}$ is a probability measure on $\mathcal{F}_{\geq s} = \sigma\{X_u;\ u \geq s\}$. Let τ_h^s denote the first passage time of X_t to the upper boundary h on $[s, \infty)$:

$$\tau_h^s = \inf\{t \geq s; X_t \geq h\}.$$

In the case where $s = t_0$, we have $\tau_h^{t_0} = \tau_h$ according to equation [1.6]. For every fixed $s \geq t_0$ and $x \in \mathbb{R}$, τ_h^s is a random variable on the probability space $(\Omega, \mathcal{F}_{\geq s}, \mathbb{P}_{s,x})$. Using a similar argument as for the distribution of τ_h

in [1.11], for $x < h$, the induced probability measure $\mathbb{P}_{s,x}^{\tau_h^s}$, that is the conditional distribution of τ_h^s under the condition $\{X_s = x\}$, has the following p.d.f.:

$$g(s,x,t,h) = \frac{h-x}{\sqrt{2\pi\sigma^2(t-s)^3}} \exp\left(-\frac{(h-x-\mu(t-s))^2}{2\sigma^2(t-s)}\right) \mathbf{1}_{\{t>s\}}. \quad [1.27]$$

If $x \geq h$, then $\mathbb{P}_{s,x}^{\tau_h^s}$ is the Dirac measure δ_s concentrated at s. For $x < h$, $z \in \mathbb{R}$ and $s < t$, writing

$$\mathbb{P}_{s,x}(X_t \leq z, \tau_h^s > t) = \mathbb{P}_{s,x}(X_t \leq z) - \mathbb{P}_{s,x}(X_t \leq z, \tau_h^s \leq t)$$

and using mostly similar arguments as for [1.7] (total probability law plus the Markov property), we get:

$$\begin{aligned}
&\mathbb{P}_{s,x}(X_t \leq z, \tau_h^s > t) \\
&= \mathbb{P}_{s,x}(X_t \leq z) - \int_s^t \mathbb{P}_{s,x}(X_t \leq z \,|\, \tau_h^s = u)\, \mathbb{P}_{s,x}^{\tau_h^s}(\mathrm{d}u) \\
&= \mathbb{P}(s,x,t,(-\infty,z]) - \int_s^t \mathbb{P}(u,h,t,(-\infty,z])\, g(s,x,u,h)\, \mathrm{d}u
\end{aligned}$$

and hence, applying Fubini's theorem and passing to densities,

$$f(s,x,t,z,h) = p(s,x,t,z) - \int_s^t p(u,h,t,z)\, g(s,x,u,h)\, \mathrm{d}u, \quad [1.28]$$

where $p(s,x,t,z)$ is the transition density of $(X_t, \mathbb{P}_{s,x})$ (namely $\mathbb{P}(s,x,t,\mathrm{d}z) = p(s,x,t,z)\,\mathrm{d}z$). From equation [1.28], the following lemma provides an explicit representation of the conditional density of a process increment under the condition that the process does not cross the level h:

$$f(s,x,t,z,h) = \frac{\partial}{\partial z}\mathbb{P}(X_t \leq z, \tau_h \notin [s,t) | X_s = x).$$

LEMMA 1.17.– Let $t_0 \leq s < t, x < h$ and $z \in \mathbb{R}$. The p.d.f. in equation [1.28] is given by

$$f(s,x,t,z,h) = \frac{1}{\sigma\sqrt{t-s}}\varphi\left(\frac{z-x-\mu(t-s)}{\sigma\sqrt{t-s}}\right) \left[1 - \exp\left(-\frac{2(h-x)(h-z)}{\sigma^2(t-s)}\right)\right] \mathbf{1}_{\{z \leq h\}}, \quad [1.29]$$

where φ is the p.d.f. of the standard normal distribution.

PROOF.– Recalling equation [1.28], we have to compute the following integral:

$$\begin{aligned} I &= \int_s^t p(u,h,t,z)\, g(s,x,u,h)\, du \\ &= \int_s^t \frac{h-x}{2\pi\sigma^2\sqrt{(t-u)(u-s)^3}} \\ &\quad \times \exp\left[-\frac{1}{2\sigma^2}\left(\frac{(z-h-\mu(t-u))^2}{t-u} + \frac{(h-x-\mu(u-s))^2}{u-s}\right)\right] du \end{aligned}$$

for $t > s \geq t_0$ and $x < h$. Applying the substitution, $y = \sqrt{(t-u)/(u-s)}$ for $s < u < t$, i.e. $u = (t+sy^2)/(1+y^2)$ with $+\infty > y > 0$, we obtain after some algebra that

$$\begin{aligned} I &= \frac{h-x}{\pi\sigma^2(t-s)} \\ &\quad \times \exp\left[-\frac{(z-h)^2 + (h-x)^2 - 2\mu(t-s)(z-x) + \mu^2(t-s)^2}{2\sigma^2(t-s)}\right] \\ &\quad \times \int_0^\infty \exp\left[-\left(\frac{\beta^2}{y^2} + \alpha^2 y^2\right)\right] dy \end{aligned}$$

with $\beta^2 = (z-h)^2/(2\sigma^2(t-s))$ and $\alpha^2 = (h-x)^2/(2\sigma^2(t-s))$. Denoting the last integral by I_1, it can be easily checked that for $\alpha \neq 0$:

$$\begin{aligned} I_1 &= \frac{\sqrt{\pi}}{4\alpha}\left[e^{2\alpha\beta}\left(\mathrm{erf}\left(\alpha z + \frac{\beta}{z}\right) - 1\right) + e^{-2\alpha\beta}\left(\mathrm{erf}\left(\alpha z - \frac{\beta}{z}\right) + 1\right)\right]\Bigg|_{z=0}^{z=\infty} \\ &= \frac{\sqrt{\pi}}{2|\alpha|}\, e^{-2|\alpha\beta|}, \end{aligned}$$

where $\text{erf}(x) = 2\pi^{-1/2}\int_0^x e^{-t^2}\,dt$ is the error function. Hence, as $x < h$ and $|\alpha| = |h-x|(2\sigma^2(t-s))^{-1/2}$, we get

$$I = \frac{1}{\sqrt{2\pi\sigma^2(t-s)}}\exp\left[-\frac{Q - 2\mu(t-s)(z-x) + \mu^2(t-s)^2}{2\sigma^2(t-s)}\right]$$

with

$$\begin{aligned} Q &= (z-h)^2 + 2|z-h|(h-x) + (h-x)^2 \\ &= \begin{cases} (z-2h+x)^2 = (z-x)^2 + 4(h-x)(h-z) & \text{if } z \leq h \\ (z-x)^2 & \text{if } z > h \end{cases} \end{aligned}$$

and, finally,

$$I = \frac{1}{\sigma\sqrt{t-s}}\varphi\left(\frac{z-x-\mu(t-s)}{\sigma\sqrt{t-s}}\right)\begin{cases}\exp\left(-\dfrac{2(h-x)(h-z)}{\sigma^2(t-s)}\right) & \text{if } z \leq h \\ 1 & \text{if } z > h\end{cases},$$

where φ is the density of the standard normal distribution. This completes the proof. □

The next lemma [KAH 10a] gives, in terms of f the density of the finite dimensional distribution of X_t starting at $x = X_s$ under the condition that h is not exceeded. Let λ^k be the k-dimensional Lebesgue measure on $\overline{\mathcal{B}}^k$ $(k \in \mathbb{N})$. We use the convention $\prod_{j=j_1}^{j_2} \alpha_j = 1$ if $j_2 < j_1$.

LEMMA 1.18.– Let $t_0 \leq s < t_1$, $\mathbf{t}_{1,k} = (t_1, \dots, t_k)$, $\mathbf{x}_{1,k} = (x_1, \dots, x_k) \in \mathbb{R}^k$ and $\mathbf{X}_{\mathbf{t}_{1,k}} = (X_{t_1}, \dots, X_{t_k})$, for $1 \leq k \leq m$. Then, the measure $\mathbb{P}_{s,x}\left(\mathbf{X}_{\mathbf{t}_{1,k}} \in \cdot, \tau_h^s \geq t_k\right)$ has the following density with respect to Lebesgue measure λ^k:

$$f_k(s, x, \mathbf{t}_{1,k}, \mathbf{x}_{1,k}, h) = f(s, x, t_1, x_1, h)\prod_{j=2}^{k} f(t_{j-1}, x_{j-1}, t_j, x_j, h). \quad [1.30]$$

PROOF.– For $k = 1$, clearly $f_1(s, x, \mathbf{t}_{1,1}, \mathbf{x}_{1,1}, h) = f(s, x, t_1, x_1, h)$ by the definition of f. For simplicity, we will prove equation [1.30] only for $k = 2$. The general case can be shown analogously by induction. Let $B_1, B_2 \in \mathcal{B}$. Then, applying the law of total probability and using the Markov property, we have

$$\begin{aligned}
&\mathbb{P}_{s,x}(\mathbf{X}_{\mathbf{t}_{1,2}} \in B_1 \times B_2,\, \tau_h^s \geq t_2) \\
&= \mathbb{P}_{s,x}(X_{t_2} \in B_2,\, \tau_h^{t_1} \geq t_2,\, X_{t_1} \in B_1,\, \tau_h^s \geq t_1) \\
&= \int_{B_1} \mathbb{P}_{s,x}(X_{t_2} \in B_2,\, \tau_h^{t_1} \geq t_2 \,|\, X_{t_1} = x_1)\, f(s,x,t_1,x_1,h)\, \mathrm{d}x_1 \\
&= \int_{B_1} \mathbb{P}_{t_1,x_1}(X_{t_2} \in B_2,\, \tau_h^{t_1} \geq t_2)\, f(s,x,t_1,x_1,h)\, \mathrm{d}x_1 \\
&= \int_{B_2 \times B_1} f(t_1,x_1,t_2,x_2,h)\, f(s,x,t_1,x_1,h)\, \mathrm{d}x_2\, \mathrm{d}x_1.
\end{aligned}$$

□

Based on these preliminary lemmas, we can derive the likelihood function $L(x_0, t_0, \mu, \sigma^2 | \mathbf{D}_{obs})$ [KAH 10a]. We shall show that $\mathbb{P}_\theta^F$ is, for all $\theta \in \Theta$, absolutely continuous with respect to the measure

$$\nu = (\lambda + \delta_{t_m}) \otimes (\lambda + \delta_\infty)^m, \qquad [1.31]$$

where λ is the Lebesgue measure on $\overline{\mathcal{B}}$ and, for every $x \in \overline{\mathbb{R}}$, δ_x is the Dirac measure on $\overline{\mathcal{B}}$ concentrated at x.

THEOREM 1.19.– Let ν be defined as in equation [1.31] on $\mathcal{B} \otimes \overline{\mathcal{B}}^{\mathrm{m}}$. Then, for all $\theta \in \Theta$, $\mathbb{P}_\theta^F$ is absolutely continuous with respect to ν and has the following Radon–Nikodym derivative:

$$\begin{aligned}
L(x_0, t_0, \mu, \sigma^2 | \mathbf{D}_{obs}) &= \frac{\mathrm{d}\mathbb{P}_\theta^F}{\mathrm{d}\nu}(\tau, \mathbf{x}) \\
= \prod_{k=1}^{m} \Bigg(g(t_{k-1}, x_{k-1}, \tau, h) &\prod_{j=1}^{k-1} f(t_{j-1}, x_{j-1}, t_j, x_j, h) \\
&\times \prod_{j=k}^{m} \mathbf{1}_{\{x_j = \infty\}} \Bigg)^{\mathbf{1}_{\{t_{k-1} \leq \tau < t_k\}}} \\
&\times \left(\prod_{j=1}^{m} f(t_{j-1}, x_{j-1}, t_j, x_j, h) \right)^{\mathbf{1}_{\{\tau = t_m\}}} \mathbf{1}_{\{t_0 \leq \tau \leq t_m\}}, \qquad [1.32]
\end{aligned}$$

where $\tau \in \mathbb{R}$, and $\mathbf{x} = (x_1, \ldots, x_m) \in \overline{\mathbb{R}}^m$.

PROOF.– Let $1 \leq i \leq j \leq n$, $(y_i, \dots, y_j) \in \overline{\mathbb{R}}^{j-i+1}$ and $t \in \mathbb{R}$. For the sake of brevity, we define $\mathbf{t}_{i,j} = (t_i, \dots, t_j)$, $\mathbf{y}_{i,j} = (y_i, \dots, y_j)$, $(-\infty, \mathbf{y}_{i,j}] = (-\infty, y_i] \times \dots \times (-\infty, y_j]$ and $E = (-\infty, t] \times (-\infty, \mathbf{y}_{1,m}]$. Furthermore, we use the abbreviation $\{\mathbf{X}_{\mathbf{t}_{i,j}} \leq \mathbf{y}_{i,j}\} = \{X_{t_i} \leq y_i, \dots, X_{t_j} \leq y_j\}$ and an analogous one for $\widetilde{\mathbf{X}}_t$. From equation [1.25], we have, for all $\theta \in \Theta$:

$$\begin{aligned}\mathbb{P}_\theta^F(E) &= \mathbb{P}_{t_0,x_0}(\min(\tau_h, t_m) \leq t, \widetilde{\mathbf{X}}_{\mathbf{t}_{1,m}} \leq \mathbf{y}_{1,m}) \\ &= \sum_{k=1}^{m} \mathbf{1}_{\{t \geq t_{k-1}\}} \mathbb{P}_{t_0,x_0}(t_{k-1} \leq \tau_h \leq \min(t, t_k), \widetilde{\mathbf{X}}_{\mathbf{t}_{1,m}} \leq \mathbf{y}_{1,m}) \\ &\qquad + \mathbf{1}_{\{t \geq t_m\}} \mathbb{P}_{t_0,x_0}(\tau_h \geq t_m, \ \widetilde{\mathbf{X}}_{\mathbf{t}_{1,m}} \leq \mathbf{y}_{1,m}). \end{aligned} \qquad [1.33]$$

To calculate the first m terms of equation [1.33], take $t_{k-1} \leq t < t_k$ with $1 \leq k \leq m$ and define $\bigcap_{k=k_1}^{k_2} A_k = \Omega$ if $k_2 < k_1$. Then, for $s \geq t_{k-1}$, by the definition of $\widetilde{\mathbf{X}}_s$:

$$\begin{aligned}&\mathbb{P}_{t_0,x_0}(t_{k-1} \leq \tau_h \leq s, \widetilde{\mathbf{X}}_{\mathbf{t}_{1,m}} \leq \mathbf{y}_{1,m}) \\ &= \mathbb{P}_{t_0,x_0}(\{t_{k-1} \leq \tau_h \leq s, \ \mathbf{X}_{\mathbf{t}_{1,k-1}} \leq \mathbf{y}_{1,k-1}\} \cap \{\widetilde{\mathbf{X}}_{\mathbf{t}_{k,m}} \leq \mathbf{y}_{k,m}\}) \\ &= \mathbb{P}_{t_0,x_0}(\tau_h^{t_{k-1}} \leq s, \ \tau_h^{t_0} \geq t_{k-1}, \mathbf{X}_{\mathbf{t}_{1,k-1}} \leq \mathbf{y}_{1,k-1}) \prod_{j=k}^{m} \mathbf{1}_{\{y_j = \infty\}}.\end{aligned}$$

Applying the law of total probability, the Markov property and lemma 1.18, it yields

$$\begin{aligned}&\mathbb{P}_{t_0,x_0}(\tau_h^{t_{k-1}} \leq s, \ \tau_h^{t_0} \geq t_{k-1}, \mathbf{X}_{\mathbf{t}_{1,k-1}} \leq \mathbf{y}_{1,k-1}) \\ &= \int_{(-\infty, \mathbf{y}_{1,k-1}]} \mathbb{P}_{t_0,x_0}(\tau_h^{t_{k-1}} \leq s \,|\, X_{t_{k-1}} = x_{k-1}) \\ &\qquad \times f_{k-1}(t_0, x_0, \mathbf{t}_{1,k-1}, \mathbf{x}_{1,k-1}, h) \lambda^{k-1}(\mathrm{d}\mathbf{x}_{1,k-1}).\end{aligned}$$

Moreover, recalling the distribution of $\mathbb{P}_{s,x}^{\tau_h^s}$, we have:

$$\begin{aligned}\mathbb{P}_{t_0,x_0}(\tau_h^{t_{k-1}} \leq s \,|\, X_{s_{k-1}} = x_{k-1}) &= \mathbb{P}_{t_{k-1},x_{k-1}}(\tau_h^{t_{k-1}} \leq s) \\ &= \int_{t_{k-1}}^{s} g(t_{k-1}, x_{k-1}, \tau, h) \, \lambda(\mathrm{d}\tau)\end{aligned}$$

if $x_{k-1} < h$. Finally, the product of indicator functions in equation [1.19] can be written as:

$$\prod_{j=k}^{m} \mathbf{1}_{\{y_j=\infty\}} = \int_{(-\infty,\mathbf{y}_{k,m}]} \prod_{j=k}^{m} \mathbf{1}_{\{x_j=\infty\}} \delta_\infty^{m-k+1}(\mathrm{d}\mathbf{x}_{k,m}).$$

Combining the above equations, we obtain:

$$\begin{aligned}
&\mathbb{P}_{t_0,x_0}(t_{k-1} \le \tau_h \le s, \widetilde{\mathbf{X}}_{\mathbf{t}_{1,m}} \le \mathbf{y}_{1,m}) \\
&= \int_{(-\infty,\mathbf{y}_{1,m}]} \int_{t_{k-1}}^{s} g(t_{k-1}, x_{k-1}, \tau, h)\, \lambda(\mathrm{d}\tau)\, f_{k-1}(t_0, x_0, \mathbf{t}_{1,k-1}, \mathbf{x}_{1,k-1}, h) \\
&\qquad \times \prod_{j=k}^{m} \mathbf{1}_{\{x_j=\infty\}} (\lambda^{k-1} \otimes \delta_\infty^{m-k+1})(\mathrm{d}\mathbf{x}_{1,m}). \qquad [1.34]
\end{aligned}$$

Now, from equations [1.29] and [1.30], it is obvious that the measure $\lambda^{k-1} \otimes \delta_\infty^{m-k+1}$ given in equation [1.34] can be replaced by:

$$(\lambda + \delta_\infty)^m = \sum_{(\mu_1,\ldots,\mu_m)\in\{\lambda,\,\delta_\infty\}^m} \bigotimes_{i=1}^{m} \mu_i$$

without altering the value of the integral, since integrating with respect to the summands of $(\lambda + \delta_\infty)^m$ gives zero except for the case $\mu_i = \lambda^{k-1} \otimes \delta_\infty^{m-k+1}$. Hence, going back to equation [1.33], using equation [1.19] and lemma 1.18, we obtain:

$$\begin{aligned}
\mathbb{P}_\theta^F(E) &= \int_{(-\infty,\mathbf{y}_{1,m}]} \Bigg(\int_{t_0}^{t} \sum_{k=1}^{m} \mathbf{1}_{\{t_{k-1}\le\tau<t_k\}} g(t_{k-1}, x_{k-1}, \tau, h)\, \lambda(\mathrm{d}\tau) \\
&\quad \times f_{k-1}(t_0, x_0, \mathbf{t}_{1,k-1}, \mathbf{x}_{1,k-1}, h) \prod_{j=k}^{m} \mathbf{1}_{\{x_j=\infty\}} \\
&\quad + \int_{t_0}^{t} \mathbf{1}_{\{\tau=t_m\}}\, \delta_{t_m}(\mathrm{d}\tau)\, f_m(t_0, x_0, \mathbf{t}_{1,m}, \mathbf{x}_{1,m}, h) \Bigg) (\lambda + \delta_\infty)^m(\mathrm{d}\mathbf{x}_{1,m}) \\
&= \int_{(-\infty,t]\times(-\infty,\mathbf{y}_{1,m}]} \Bigg(\sum_{k=1}^{m} \mathbf{1}_{\{t_{k-1}\le\tau<t_k\}} g(t_{k-1}, x_{k-1}, \tau, h)
\end{aligned}$$

$$\times \prod_{j=1}^{k-1} f(t_{j-1}, x_{j-1}, t_j, x_j, h) \prod_{j=k}^{m} \mathbf{1}_{\{x_j=\infty\}}$$

$$+ \mathbf{1}_{\{\tau=t_m\}} \prod_{j=1}^{m} f(t_{j-1}, x_{j-1}, t_j, x_j, h) \Bigg) \nu(\mathrm{d}(\tau, \mathbf{x}_{1,m})),$$

where $\nu = (\lambda + \delta_{t_m}) \otimes (\lambda + \delta_\infty)^m$. Thus, the integrand of the last equation is a version of the Radon–Nikodym derivative of $\mathbb{P}_\theta^F$ with respect to ν, since $\mathcal{B} \otimes \overline{\mathcal{B}}^{\mathrm{m}}$ is generated by the sets E. To complete the proof note that this integrand and the right-hand side of equation [1.32] are equivalent. □

In general, we observe not only one single degradation process, but n i.i.d. degradation processes $X^{(1)}, \ldots, X^{(n)}$ corresponding to n independent and identical items. Let $t_{i1}, \ldots, t_{im_i}$ with $t_0 < t_{i1} < \ldots < t_{im_i} < \infty$ be the m_i fixed observation times for $X^{(i)}$, $M = \sum_{i=1}^n m_i$ and $\tau_h^{(i)}$, the lifetime associated with $X^{(i)}$. In this case, the censored observations have the form:

$$\begin{aligned}&(\tau_i, x_{i1}, \ldots, x_{im_i})\\ =&(\min(\tau_h^{(i)}, t_{im_i}), G_{t_{i1}}(\tau_h^{(i)}, X^{(i)}(t_{i1})), \ldots, G_{t_{im_i}}(\tau_h^{(i)}, X^{(i)}(t_{im_i}))).\end{aligned}$$

Then, the statistical model is given by:

$$(\mathbb{R}^n \times \overline{\mathbb{R}}^M, \mathcal{B}^n \otimes \overline{\mathcal{B}}^M, (\otimes_{i=1}^n \mathbb{P}_\theta^F)_{\theta=(\mu,\sigma^2,x_0,t_0,h)\in\Theta\subset\mathbb{R}^5})$$

with the following likelihood function:

$$L(x_0, t_0, \mu, \sigma^2|\mathbf{D}_{obs}) = \prod_{i=1}^n L^{(i)}(x_0, t_0, \mu, \sigma^2|\mathbf{D}_{obs}) \ ,$$

where $\tau = (\tau_1, \ldots, \tau_n) \in \mathbb{R}^n$, $\mathbf{x}^{(i)} = (x_{i1}, \ldots, x_{im_i}) \in \overline{\mathbb{R}}^{m_i}$, $i \in \{1, \ldots, n\}$ and $L^{(i)}$ is the likelihood function of the i-th observed process.

EXAMPLE 1.20.– We consider degradation processes with $x_0 = 0$ and $t_0 = 0$. Substituting equations [1.27] and [1.29] for the functions g and f in equation [1.32], we get the likelihood function of the i-th observed process for $i \in \{1, \ldots, n\}$,

$$L(\mu, \sigma^2, h|\mathbf{D}_{obs}) =$$

$$\prod_{j=1}^{k-1} \left(\frac{1}{\sqrt{\sigma^2(t_{i,j} - t_{i,j-1})}} \varphi\left(\frac{(x_{i,j} - x_{i,j-1}) - \mu(t_{i,j} - t_{i,j-1})}{\sqrt{\sigma^2(t_{i,j} - t_{i,j-1})}} \right) \right.$$

$$\left. \times \left[1 - \exp\left(-\frac{2(h - x_{i,j-1})(h - x_{i,j})}{\sigma^2(t_{i,j} - t_{i,j-1})} \right) \right] \right) \prod_{j=k}^{m_i} \mathbf{1}_{\{x_j = \infty\}}$$

$$\times \frac{h - x_{i,k-1}}{\sqrt{\sigma^2(\tau_i - t_{i,k-1})^3}} \varphi\left(\frac{h - x_{i,k-1} - \mu(\tau_i - t_{i,k-1})}{\sqrt{\sigma^2(\tau_i - t_{i,k-1})}} \right), \quad [1.35]$$

if $t_{k-1} \leq \tau_i < t_k$ and $1 \leq k \leq m_i$ and

$$L(\mu, \sigma^2, h|\mathbf{D}_{obs}) =$$

$$\prod_{j=1}^{m_i} \left(\frac{1}{\sqrt{\sigma^2(t_{i,j} - t_{i,j-1})}} \varphi\left(\frac{(x_{i,j} - x_{i,j-1}) - \mu(t_{i,j} - t_{i,j-1})}{\sqrt{\sigma^2(t_{i,j} - t_{i,j-1})}} \right) \right.$$

$$\left. \times \left[1 - \exp\left(-\frac{2(h - x_{i,j-1})(h - x_{i,j})}{\sigma^2(t_{i,j} - t_{i,j-1})} \right) \right] \right), \quad [1.36]$$

if $\tau_i \geq t_{m_i}$. To compare our estimates based on the likelihood function given by equation [1.35] and by equation [1.36] with previous results, we consider the case where only degradation process increments are observed. As shown in section 1.6.1, in this case, the likelihood function of the i-th observed process is given by:

$$L_D(\mu, \sigma^2|\mathbf{D}_{obs})$$

$$= \prod_{j=1}^{m_i} \frac{1}{\sqrt{\sigma^2(t_{i,j} - t_{i,j-1})}} \varphi\left(\frac{(x_{i,j} - x_{i,j-1}) - \mu(t_{i,j} - t_{i,j-1})}{\sqrt{\sigma^2(t_{i,j} - t_{i,j-1})}} \right). \quad [1.37]$$

We notice that the likelihood function given by equation [1.37] can be obtained from equation [1.36] by making h tend to infinity, which means that the boundary level h does not exist and, consequently, exceeding the boundary is impossible.

As an example, we have simulated a sample of $n = 5$ Wiener processes with $m_i = 10$ observation times $t_{i,j}$ for each process, which have equal distances $t_{i,j} - t_{i,j-1} = 1$ and with true parameter values $\mu = 1.0$, $\sigma^2 = 0.25$, $t_0 = 0$, $x_0 = 0$ and $h = 5.0$. The simulated data are shown in Table 1.2.

i	1	2	3	4	5
$x_{i,j} - x_{i,j-1}$	1.290	0.397	0.744	0.984	1.223
	1.096	1.619	1.125	0.904	0.769
	0.715	0.427	1.047	1.384	1.775
	0.339	1.574	0.530	1.000	0.567
	0.028		0.123		0.426
	1.231		1.346		
τ_i	6.559	4.967	6.018	4.461	5.138

Table 1.2. *A simulated sample*

Supposing that the parameter values $t_0 = 0$ and $x_0 = 0$ are known, we first estimate the parameters μ and σ^2 maximizing the likelihood function given by equation [1.4], which means that we ignore that the degradation process does not exceed the boundary level between observation times. Using the maximum likelihood estimators described in section 1.6.1, we get:

$$\widehat{\mu}_D = 0.9198, \quad \text{and} \quad \widehat{\sigma}^2_D = 0.1983.$$

In Figure 1.10, 95 and 99% confidence regions for μ and σ^2 based on the asymptotic χ^2 distribution (with two degrees of freedom) of the likelihood ratio are shown for this model.

Now, supposing again that $t_0 = 0$ and $x_0 = 0$ are known, we compute the maximum likelihood estimates for the parameters μ, σ^2 and h based on the likelihood function given by equation [1.35] of the exact model. The maximization of the likelihood function yields the point estimates

$$\widehat{\mu} = 0.9222, \quad \widehat{\sigma^2} = 0.1915 \quad \text{and} \quad \widehat{h} = 5.0047.$$

The numerical optimization procedure turned out to be difficult to converge, so that a good initial value of the parameters has to be known. Such initial values may be the estimates of μ and σ^2 from the likelihood function given by equation [1.4] and an initial value of h can be found from the likelihood function based on the inverse Gaussian density (see equation [1.11]). Figures 1.11– 1.13 show 0.95- and 0.99-confidence regions based on the asymptotic χ^2_2 distribution of the likelihood ratio for the parameters (μ, σ^2), (μ, h) and (σ^2, h), respectively.

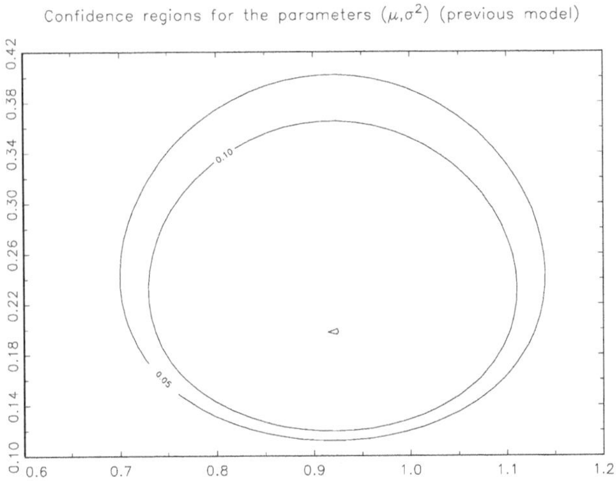

Figure 1.10. *Confidence estimates for* (μ, σ^2)

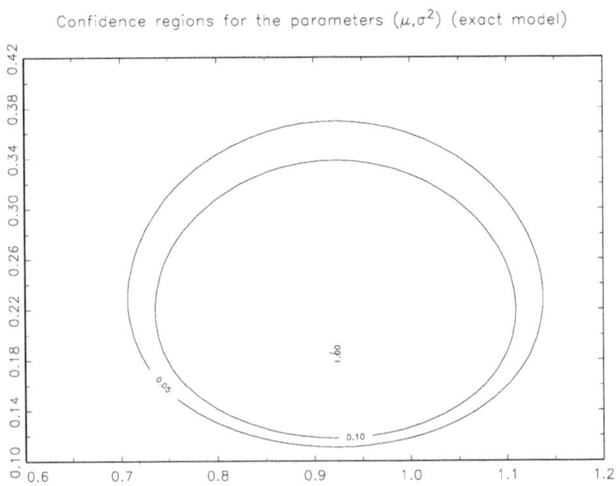

Figure 1.11. *Confidence estimates for* (μ, σ^2)

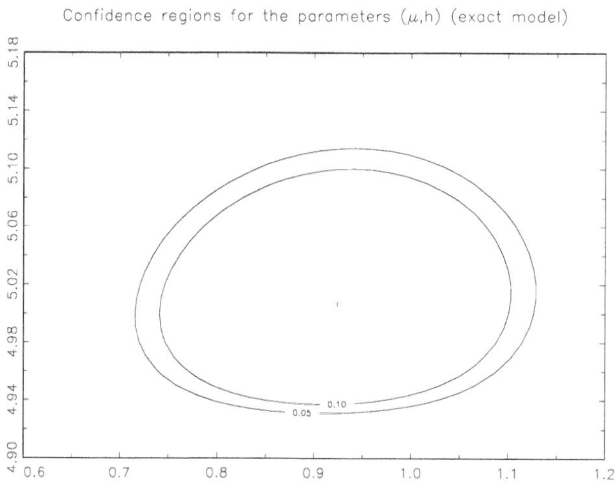

Figure 1.12. *Confidence estimates for* (μ, h)

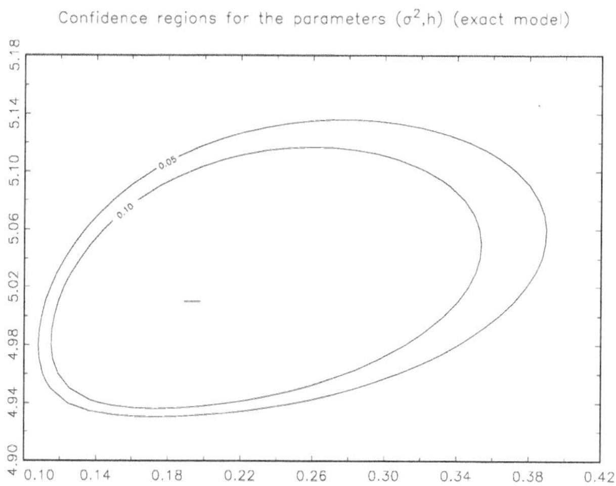

Figure 1.13. *Confidence estimates for* (σ^2, h)

Note that, in this case, it is also possible to estimate parameter h.

In an intensive further simulation study, we have estimated the partial parameter vector $\theta^* = (\mu, \sigma^2, h)$, whereas t_0 and x_0 were assumed to be known. Given an observation vector $X = (X_1, \ldots, X_n)$ with $X^{(i)} = (\min(\tau_i, t_{i,m_i}), x_{i,1}, \ldots, x_{i,M_i})$, only the maximum likelihood estimator for the drift parameter μ from the likelihood functions defined by equations [1.35] and [1.36] can be explicitly obtained as follows:

$$\widehat{\mu} = \frac{\sum_{i=1}^{n} \left(x_{i,M_i} - x_{i,0} + \mathbf{1}_{\{\tau_i < t_{i,m_i}\}} (\widehat{h} - x_{i,M_i}) \right)}{\sum_{i=1}^{n} \left(t_{i,M_i} - t_{i,0} + \mathbf{1}_{\{\tau_i < t_{i,m_i}\}} (\tau_i - t_{i,M_i}) \right)},$$

where $M_i = \max\{k : t_{i,k} < \tau_h^{(i)}\}$. The estimators $\widehat{\sigma^2}$ and $\widehat{h}$ were determined numerically.

Table 1.3 summarizes the estimated mean-squared errors of the maximum likelihood estimators $\widehat{\mu}$, $\widehat{\sigma^2}$ and $\widehat{h}$ from 1,000 simulation runs and compares them with those of the estimators $\widehat{\mu}_D$ and $\widehat{\sigma^2}_D$ from the likelihood function given by equation [1.4] that served as initial values for the optimization procedure. In each simulation run, $n = 50$ Wiener processes were simulated with $m_i = 10$ observation times $t_{i,j}$ and equal distances $t_{i,j} - t_{i,j-1} = 1$. The true parameter values were $\mu = 5$, $\sigma^2 \in \{1, 2, 5\}$, $t_0 = 0$, $x_0 = 0$ and $h \in \{20, 40, 50\}$. Additionally, Table 1.3 gives the average number $\bar{d}$ of observation times $t_{i,j}$ per degradation process before a failure.

Table 1.3 shows that, for all parameter choices in the simulation study, the mean-squared errors of the estimators $\widehat{\mu}$, $\widehat{\sigma^2}$ and $\widehat{h}$ based on the likelihood functions defined by equations [1.35] and [1.36] are smaller than those of the estimators $\widehat{\mu}_D$ and $\widehat{\sigma^2}_D$ based on the likelihood function in equation [1.4]. In particular, the general model based on the observation of degradation increments and failure times provides better estimates of the process parameters μ and σ^2, especially if nearly all items have failed during the experiment and if the number of actually observed degradation levels is small in comparison with the number of planned observation times.

μ	σ^2	h	$\bar{d}$	Empirical mean squared errors				
				$\widehat{\mu}$	$\widehat{\mu}_D$	$\widehat{\sigma^2}$	$\widehat{\sigma^2}_D$	$\widehat{h}$
		20	3.50	0.0107	0.0198	0.0347	0.0440	0.0028
5	2	40	7.55	0.0046	0.0062	0.0191	0.0196	0.0027
		50	9.34	0.0041	0.0042	0.0161	0.0172	0.0103
	1		3.51	0.0046	0.0085	0.0093	0.0114	0.0015
5	2	20	3.50	0.0107	0.0198	0.0347	0.0440	0.0028
	5		3.45	0.0358	0.0454	0.2992	0.5238	0.0043

Table 1.3. *Mean squared errors of maximum likelihood estimators in 1,000 simulation runs*

1.7. Extension and related models

Although it is widely used as a degradation model, the Wiener process suffers of some limitation and so it cannot be used anytime. One major drawback is that the average degradation is linear over time (see equation [1.3]). Some extensions or related models have hence been proposed in the literature, to remove this limitation. We here review two of them. With that aim, we introduce a time-change function $A : \mathbb{R}^+ \longrightarrow \mathbb{R}^+$ that is assumed to be increasing, continuous and such that $A(0) = 0$ and $\lim_{t\to\infty} A(t) = \infty$.

1.7.1. *Time-scaled Wiener processes*

A first extension consists of changing the time scale. More precisely, starting from a Wiener process $(Y_t)_{t\geq 0}$ with unit mean:

$$\forall t \geq 0, \quad Y_t = t + \sigma B_t,$$

a time-scaled Wiener process is defined as follows:

$$\forall t \geq 0, \quad X_t = Y_{A(t)} = A(t) + \sigma B_{A(t)}.$$

It is easy to check that, for any $t \geq 0$, X_t has the Gaussian distribution with mean $A(t)$ and variance $\sigma^2 A(t)$.

The non-overlapping increments of such a model are still independent but they are not stationary any more. Hence, the process remains additive (but it is no longer a Lévy process). Anyway, the likelihood function can easily be written based on the increments and maximized for a parametric form of A. A semi-parametric estimator has also been proposed by Wang [WAN 09b] based on a profile pseudo-log-likelihood approach (just as he did for the gamma process, see section 2.8.1 for details). Note that Wang has also considered the case of random effects for this model: in [WAN 09b], he has assumed that A is random and inverse Gaussian distributed, while in [WAN 10], he has considered a random drift and volatility with extra parameters to be estimated (hierarchical parameters).

The distribution of the time to failure can easily be expressed with respect to the inverse Gaussian distribution. Indeed, setting $\tau_{h,X}$ and $\tau_{h,Y}$ to be the time to failures associated with $(X_t)_{t\geq 0}$ and $(Y_t)_{t\geq 0}$, respectively, it is readily seen that $\tau_{X,h} = A^{-1}(\tau_{Y,h})$, where A^{-1} is the inverse function of A. Recalling that $\tau_{Y,h}$ has the inverse Gaussian distribution with parameters $(\beta = h^2/\sigma^2, \mu_1 = h)$ and using the expression of its p.d.f. given by equation [1.11], we have:

$$\forall t \geq 0, \quad f_{\tau_{X,h}}(t) = \frac{hA'(t)}{\sqrt{2\pi\sigma^2 A(t)^3}} \exp\left(-\frac{(h - A(t))^2}{2\sigma^2 A(t)}\right).$$

The mean time to failure may be quite difficult to compute explicitly, except for some particular cases (e.g. for a power transformation $A(t) = at^\gamma$).

1.7.2. *Brownian motion with nonlinear drift*

Instead of using a time-scaling, only the linear drift μt of the Wiener process can be replaced in equation [1.3] by the time-change $A(t)$. Hence, the model turns out to be:

$$\forall t \geq 0, \quad X_t = A(t) + \sigma B_t.$$

For any $t \geq 0$, X_t has the Gaussian distribution with mean $A(t)$, but with variance $\sigma^2 t$. Here again, the process remains additive, but it is not a Lévy process. The likelihood function can easily be written based on the independence of the increments. This model has been studied, for instance, by Si *et al.* [SI 12] (see also [WAN 14]).

Here, the distribution of the time to failure cannot be expressed explicitly for any choice of the function A. This can be viewed as the problem of determining the first passage time distribution of a time-dependent boundary by a Brownian motion (see [WAN 97], for instance). For any general boundary, some numerical approximations have been proposed in the literature. For instance, Wang and Pötzelberger [WAN 07] have proposed a computational approach based on a linear piecewise approximation for A and using Monte Carlo simulations.

2

Gamma Processes

As already seen in Chapter 1, the Wiener process models a continuous deterioration with an increasing trend. However, it has non-monotonous trajectories and this is one major drawback of this model for the analysis of degradation measurements. Consequently, the gamma process is sometimes preferred to the Wiener process. As we will see, it is a pure jump process with an infinite number of jumps over any finite time interval. Hence, the gamma process is well suited to model some non-decreasing deterioration that accumulates over time through many tiny increments.

When not further specified, all random variables and stochastic processes are defined on a common probability space $(\Omega, \mathcal{A}, \mathbb{P})$.

2.1. Gamma distribution

We recall here a few basic facts on the gamma distribution.

DEFINITION 2.1.– *Let $a, b > 0$. A random variable X is said to be gamma distributed with shape parameter a and rate parameter b (parameters (a, b) for short) if its distribution admits the following probability density function with respect to Lebesgue measure:*

$$f_{a,b}(x) = \frac{b^a}{\Gamma(a)} x^{a-1} e^{-bx} \mathbf{1}_{\mathbb{R}_+}(x), \text{ for all } x \in \mathbb{R},$$

where

$$\Gamma(a) = \int_0^{+\infty} x^{a-1} e^{-x} \, \mathrm{d}x$$

is the gamma function.

The gamma distribution with parameters (a, b) is denoted by $\mathcal{G}(a, b)$ in the following, and its p.d.f., c.d.f. and survival function by $f_{a,b}$, $F_{a,b}$ and $\bar{F}_{a,b}$, respectively. Note that another standard parameterization in the literature is (a, β) with $\beta = \frac{1}{b}$. In that case, parameter β is called *scale* parameter.

In the following, we will allow the shape parameter a to be zero and the gamma distribution $\mathcal{G}(0, b)$ will refer to the Dirac mass at point 0, namely to the distribution of the constant r.v. equal to 0.

A random variable X with gamma distribution is non-negative and its distribution is characterized by its Laplace transform $\mathcal{L}_X$, where we recall that, for any non-negative r.v. X with p.d.f. f_X, we have:

$$\mathcal{L}_X(s) = \mathbb{E}\left(e^{-sX}\right) = \int_0^{+\infty} e^{-sx} f_X(x)\, \mathrm{d}x, \text{ for all } s \geq 0.$$

PROPOSITION 2.2.– Let X be gamma distributed $\mathcal{G}(a, b)$, with $a, b > 0$. Then, its Laplace transform is:

$$\mathcal{L}_X(s) = \left(\frac{b}{b+s}\right)^a, \text{ for all } s \geq 0.$$

Also:

$$\mathbb{E}(X) = \frac{a}{b} \quad \text{and} \quad \mathrm{Var}(X) = \frac{a}{b^2}.$$

Note that the previous results are also valid for $a = 0$.

PROOF.– When $a > 0$, we have:

$$\begin{aligned} \mathcal{L}_X(s) &= \int_0^{+\infty} e^{-sx} \frac{b^a}{\Gamma(a)} x^{a-1} e^{-bx}\, \mathrm{d}x \\ &= \frac{b^a}{(b+s)^a} \int_0^{+\infty} f_{a,b+s}(x)\, \mathrm{d}x = \left(\frac{b}{b+s}\right)^a, \end{aligned}$$

where $f_{a,b+s}$ is the p.d.f. of $\mathcal{G}(a, b+s)$. We easily derive $\mathbb{E}(X)$ and $\mathrm{Var}(X)$ using that $\mathbb{E}(X) = -\mathcal{L}'_X(0^+)$ and $\mathbb{E}\left(X^2\right) = \mathcal{L}''_X(0^+)$. □

COROLLARY 2.3.– Let X_1, ..., X_n be independent gamma distributed r.v.s with parameters (a_1, b), ..., (a_n, b), respectively (same rate parameter). Then, $\sum_{i=1}^{n} X_i$ is gamma distributed $\mathcal{G}\left(\sum_{i=1}^{n} a_i, b\right)$.

PROOF.– Based on the independence, we have:

$$\mathcal{L}_{\sum_{i=1}^{n} X_i}(s) = \mathbb{E}\left(e^{-s\sum_{i=1}^{n} X_i}\right) = \prod_{i=1}^{n} \mathbb{E}\left(e^{-sX_i}\right)$$
$$= \prod_{i=1}^{n} \left(\frac{b}{b+s}\right)^{a_i} = \left(\frac{b}{b+s}\right)^{\sum_{i=1}^{n} a_i}$$

where the last expression is the Laplace transform of $\mathcal{G}\left(\sum_{i=1}^{n} a_i, b\right)$. □

As the gamma distribution $\mathcal{G}(1, \lambda)$ coincides with the exponential distribution $\mathcal{E}(\lambda)$ with mean $\frac{1}{\lambda}$, we directly get the following corollary.

COROLLARY 2.4.– Let X_1, ..., X_n be independent and identically exponentially distributed r.v.s with mean $\frac{1}{\lambda}$. Then, $\sum_{i=1}^{n} X_i$ is gamma distributed $\mathcal{G}(n, \lambda)$ with p.d.f.

$$f(x) = \frac{\lambda^n}{(n-1)!} x^{n-1} e^{-\lambda x} \mathbf{1}_{\mathbb{R}_+}(x).$$

We can also derive another basic fact from proposition 2.2.

COROLLARY 2.5.– Let X be gamma distributed $\mathcal{G}(a, b)$ and let $c > 0$. Then, cX is gamma distributed $\mathcal{G}\left(a, \frac{b}{c}\right)$.

PROOF.– For all $s \geq 0$:

$$\mathcal{L}_{cX}(s) = \mathbb{E}\left(e^{-scX}\right) = \left(\frac{b}{b+sc}\right)^a = \left(\frac{b/c}{b/c+s}\right)^a$$

which provides the result. □

Recall that the beta distribution $\mathcal{B}(a_1, a_2)$ with parameters $a_1, a_2 > 0$ admits the following p.d.f. with respect of Lebesgue measure:

$$f(x) = \frac{x^{a_1}(1-x)^{a_2}}{B(a_1, a_2)} \mathbf{1}_{[0,1]}(x),$$

where

$$B(a_1, a_2) = \frac{\Gamma(a_1 + a_2)}{\Gamma(a_1)\Gamma(a_2)}$$

is the beta function. We then have the following result.

PROPOSITION 2.6.– Let X and Y be two random variables and let $a_1, a_2, b > 0$. Then, the two following points are equivalent:

1) $X + Y$ and $\frac{X}{X+Y}$ are independent, with respective distributions $\mathcal{G}(a_1 + a_2, b)$ and $B(a_1, a_2)$;

2) X and Y are independent with respective distributions $\mathcal{G}(a_1, b)$ and $\mathcal{G}(a_2, b)$.

PROOF.– Let $(Z, U) = \left(X + Y, \frac{X}{X+Y}\right)$, so that $(X, Y) = (ZU, Z(1 - U))$. We set φ to be any measurable non-negative function. Assume point 1 to be true. Then:

$$\mathbb{E}(\varphi(X, Y)) = \mathbb{E}(\varphi(ZU, Z(1 - U)))$$

$$= cst \times \int_0^{+\infty} \int_0^1 \varphi(zu, z(1 - u))\, z^{a_1 + a_2 - 1} e^{-bz} u^{a_1 - 1} (1 - u)^{a_2 - 1}\, du\, dz$$

$$= cst \times \int_0^{+\infty} \int_0^{+\infty} \varphi(x, y)(x + y)^{a_1 + a_2 - 2} e^{-b(x+y)} \times \left(\frac{x}{x+y}\right)^{a_1 - 1} \left(\frac{y}{x+y}\right)^{a_2 - 1} dx\, dy$$

$$= cst \times \int_0^{+\infty} \int_0^{+\infty} \varphi(x, y)\, e^{-b(x+y)} x^{a_1 - 1} y^{a_2 - 1} dx\, dy$$

where cst stands for a constant and where we set $x = zu$ and $y = z(1 - u)$ (with $du\, dz = \frac{1}{x+y}\, dx\, dy$). This shows that the joint p.d.f. of (X, Y) is of the form

$$cst \times e^{-b(x+y)} x^{a_1 - 1} y^{a_2 - 1} = cst \times e^{-bx} x^{a_1 - 1} \times e^{-by} y^{a_2 - 1}$$

and point 2 is true.

Conversely, assume point 2 to be true. Then:

$$\begin{aligned}\mathbb{E}\left(\varphi\left(Z,U\right)\right) &= \mathbb{E}\left(\varphi\left(X+Y,\frac{X}{X+Y}\right)\right)\\ &= cst \times \int_0^{+\infty}\int_0^{+\infty} \varphi\left(x+y,\frac{x}{x+y}\right) e^{-b(x+y)} x^{a_1-1} y^{a_2-1} \mathrm{d}x\,\mathrm{d}y\\ &= cst \times \int_0^{+\infty}\int_0^{1} \varphi\left(z,u\right) e^{-bz} z^{a_1+a_2-1} u^{a_1-1}\left(1-u\right)^{a_2-1} \mathrm{d}u\,\mathrm{d}z\end{aligned}$$

where $(z,u) = \left(x+y,\frac{x}{x+y}\right)$ (with $\mathrm{d}x\,\mathrm{d}y = z\,\mathrm{d}u\,\mathrm{d}z$), which shows the result.

□

We now provide some stochastic comparison results between gamma distributed r.v.s and we first recall the definition of several stochastic orders.

DEFINITION 2.7.– *Let X and Y be two random variables with c.d.f. F_X and F_Y and survival functions $\bar{F}_X$ and $\bar{F}_Y$, respectively.*

The r.v. X is said to be smaller than the r.v. Y with respect to the usual stochastic order ($X \prec_{sto} Y$) if $\bar{F}_X \leq \bar{F}_Y$.

The r.v. X is said to be smaller than the r.v. Y with respect to the hazard rate order ($X \prec_{hr} Y$) if

$$\bar{F}_X\left(y\right)\bar{F}_Y\left(x\right) \leq \bar{F}_X\left(x\right)\bar{F}_Y\left(y\right)$$

for all $y \geq x$.

The r.v. X is said to be smaller than the r.v. Y with respect to the reversed hazard rate order ($X \prec_{rhr} Y$) if

$$F_X\left(y\right)F_Y\left(x\right) \leq F_X\left(x\right)F_Y\left(y\right)$$

for all $y \geq x$.

Assume that the distributions of X and Y are absolutely continuous with respect to Lebesgue measure with p.d.f. f_X and f_Y, respectively. Then, the r.v. X is said to be smaller than the r.v. Y with respect to the likelihood ratio order if

$$f_X\left(y\right)f_Y\left(x\right) \leq f_X\left(x\right)f_Y\left(y\right)$$

for all $y \geq x$.

We recall that the likelihood ratio order implies both hazard and reversed hazard rate orders, which themselves imply the usual stochastic order, see [SHA 07] for more details on these notions.

PROPOSITION 2.8.– Let X and Y be gamma distributed r.v.s with distributions $\mathcal{G}(a,b)$ and $\mathcal{G}(a',b)$ (same rate parameter), respectively, where $a \leq a'$. Then, X is smaller than Y with respect to the likelihood ratio order, and consequently also with respect to both hazard and reversed hazard rate orders and to the usual stochastic order.

PROOF.– Let $x \leq y$. We have:

$$f_X(x) f_Y(y) - f_X(y) f_Y(x) = \frac{b^{a'}}{\Gamma(a')} \frac{b^a}{\Gamma(a)} e^{-b(x+y)} (xy)^{a-1} \left(y^{a'-a} - x^{a'-a}\right) \geq 0$$

because $a \leq a'$. □

2.2. Poisson random measures

For sake of completeness, we here introduce Poisson random measures, which are used in section 2.4 to construct series representations of gamma processes. Note that we do not intend to make a general presentation of (Poisson) random measures and we mostly restrict ourselves to the specific setting of purely atomic random measures on $\left(\mathbb{R}_+^d, \mathcal{B}\left(\mathbb{R}_+^d\right)\right)$, which is sufficient for our applicative purposes, where we only need purely atomic random measures on $\mathbb{R}_+^2$ and $\mathbb{R}_+^3$. The interested reader may for example refer to [ÇIN 11, JAC 87, NEV 77] for a more detailed account. *This section can be skipped in the first reading.*

We first recall some elementary vocabulary. Let $\lambda > 0$. A random variable X is said to be Poisson distributed with parameter λ (denoted by $X \sim \mathcal{P}(\lambda)$) if

$$\mathbb{P}(X = k) = e^{-\lambda} \frac{\lambda^k}{k!}, \text{ for all } k \in \mathbb{N}.$$

The moments of a Poisson distributed r.v. X with parameter λ are $\mathbb{E}(X) = \text{Var}(X) = \lambda$ and its Laplace transform is

$$\mathcal{L}_X(s) = \mathbb{E}\left(e^{-sX}\right) = \sum_{k=0}^{+\infty} e^{-sk} e^{-\lambda} \frac{\lambda^k}{k!} = e^{-\lambda\left(1-e^{-s}\right)}, \text{ for all } s \geq 0. \quad [2.1]$$

By extension, a Poisson distributed r.v. with parameter $\lambda = 0$ is null a.s.; in case of an infinite mean ($\lambda = +\infty$), it is equal to $+\infty$ a.s.

A process $(N_t)_{t\geq 0}$ is called a *counting process* if there exists a sequence of almost surely distinct random variables $(T_n)_{n\geq 1}$ such that

$$N_t = \sum_{n=1}^{+\infty} \mathbf{1}_{[0,t]}(T_n) = \sum_{n=1}^{+\infty} \delta_{T_n}([0,t]), \text{ for all } t \geq 0,$$

where δ_a is the Dirac mass at point $a \in \mathbb{R}_+$.

DEFINITION 2.9.– *Let $\Lambda(t)$ be a continuous and non-decreasing function such that $\Lambda(0) = 0$. A counting process $(N_t)_{t\geq 0}$ is called a (non-homogeneous) Poisson process with mean rate function $\Lambda(t)$ if $(N_t)_{t\geq 0}$ has independent increments and is such that $N_t - N_s$ is Poisson distributed $\mathcal{P}(\Lambda(t) - \Lambda(s))$, for all $0 \leq s \leq t$. In the specific case where $\Lambda(t) = \lambda t$ for all $t \geq 0$, the Poisson process is said to be homogeneous with rate λ.*

A homogeneous Poisson process has stationary increments (see, for instance, [ROS 96]) and it is a Lévy process, whereas a non-homogeneous Poisson process is an additive process (see definition I.1).

The points of a Poisson process are easy to simulate on some time interval $[0,T]$, using the fact that N_T is Poisson distributed $\mathcal{P}(\Lambda(T))$ and that given that $N_T = n$ with $n \in \mathbb{N}^*$, the random vector $(T_1, \cdots, T_n)$ is identically distributed as the order statistics $\left(X_{(1)}, \cdots, X_{(n)}\right)$ of some i.i.d. random variables $X_1, \cdots, X_n$ with c.d.f. $\frac{\Lambda(t)}{\Lambda(T)}$ on $[0,T]$. Assuming Λ to be one-to-one from $[0,T]$ into $[0,\Lambda(T)]$ and using the inverse c.d.f. method for simulating $X_1, \cdots, X_n$ (or considering the left-continuous pseudo-inverse in case of a non-one-to-one function, see [ASM 07, Prop. 2.2] for instance), we get the following algorithm.

ALGORITHM 2.10.– [Simulation of an observation $(t_i)_{1\leq i\leq n}$ of the points of $(N_t)_{t\geq 0}$ on $[0,T]$, sorted by increasing order]

– generate an observation n of N_T according to $\mathcal{P}(\Lambda(T))$;

– if $n \geq 1$;

- generate (independently) n i.i.d. observations $u_1, \cdots, u_n$ according to $\mathcal{U}([0,1])$,

- set $x_i = \Lambda^{-1}(\Lambda(T)\ u_i)$ for all $1 \leq i \leq n$,

- sort the x_i's by increasing order and set $t_i = x_{(i)}$ (i-th value among the x_j's) for all $1 \leq i \leq n$.

In the special case of a homogeneous Poisson process ($\Lambda(t) = \lambda t$), we simply set $x_i = T\ u_i$ (the X_i's are uniformly distributed $\mathcal{U}([0,T])$).

Another possibility is to generate the points $T_1, \cdots, T_n, \cdots$ sequentially, using the fact that given $T_n = t_n$ (with $n \geq 0$), $T_{n+1} - T_n$ has the following conditional c.d.f.: $\mathbb{P}(T_{n+1} - T_n \leq t | T_n = t_n) = 1 - e^{-(\Lambda(t+t_n) - \Lambda(t_n))}$ and next using the inverse c.d.f. method (assuming Λ to be one-to-one). This provides the following algorithm.

ALGORITHM 2.11.– Sequential simulation of an observation $(t_i)_{1 \leq i \leq n}$ of the points of $(N_t)_{t \geq 0}$ on $[0, T]$, sorted by increasing order

– set $i = 0$, $t_i = 0$;

– while $t_i < T$ do:

- generate (independently) an observation u_{i+1} according to $\mathcal{U}([0,1])$;

- set $y_{i+1} = \Lambda^{-1}(\log(u_{i+1}))$ (observation of $T_{i+1} - T_i$);

- set $t_{i+1} = t_i + y_{i+1}$;

- set $i = i + 1$.

In the special case of a homogeneous Poisson process ($\Lambda(t) = \lambda t$), we simply set $y_{i+1} = -\frac{\log(u_{i+1})}{\lambda}$ (because the $T_{i+1} - T_i$'s are exponentially distributed $\mathcal{E}(\lambda)$).

Note that when the sorting step of algorithm 2.10 is unnecessary (as will be the case in section 2.5.3 for the simulation of gamma processes through compound Poisson approximations), algorithm 2.10 is more efficient, because it does not contain any test whereas the while loop of algorithm 2.11. requires a test at each step. When there is a need for the points to be ordered, the comparison between the efficiency of the two algorithms is not so clear.

A measure μ on $(\mathbb{R}_+^d, \mathcal{B}(\mathbb{R}_+^d))$ is said to be *atomic* if it is of the shape

$$\mu = \sum_{n=1}^{\infty} \delta_{\mathbf{x}_n}$$

where the $\mathbf{x}_n$'s belong to $\mathbb{R}_+^d$ (not necessarily distinct) and if the sum is locally finite, namely such that $\mu(B) < +\infty$ for all compact Borel set B [NEV 77]. The $\mathbf{x}_n$'s are called the *atoms* (or *points*) of μ. When the $\mathbf{x}_n$'s are distinct, then $\mu(\{\mathbf{x}_n\}) = 1$ for each atom $\mathbf{x}_n$ and the measure μ is said to be *purely atomic*. (The measure μ is also called a *counting* measure).

A measure μ on $\left(\mathbb{R}_+^d, \mathcal{B}\left(\mathbb{R}_+^d\right)\right)$ is said to be *diffuse* if $\mu(\{\mathbf{x}\}) = 0$ for all $\mathbf{x} \in \mathbb{R}_+^d$. Any measure which is absolutely continuous with respect of Lebesgue measure is diffuse.

We are now ready to define random measures.

DEFINITION 2.12.– *Let M be a function $M : \Omega \times \mathcal{B}\left(\mathbb{R}_+^d\right) \longrightarrow \bar{\mathbb{R}}_+$. The function M is a random measure on $\left(\mathbb{R}_+^d, \mathcal{B}\left(\mathbb{R}_+^d\right)\right)$ (or on $\mathbb{R}_+^d$ for short) if:*

1) *for almost every $\omega \in \Omega$, the function $\bar{M}_\omega : \mathcal{B}\left(\mathbb{R}_+^d\right) \longrightarrow \bar{\mathbb{R}}_+$ such that $\bar{M}_\omega(A) = M(\omega, A)$ for all $A \in \mathcal{B}\left(\mathbb{R}_+^d\right)$ is a measure on $\left(\mathbb{R}_+^d, \mathcal{B}\left(\mathbb{R}_+^d\right)\right)$,*

2) *for each $A \in \mathcal{B}\left(\mathbb{R}_+^d\right)$, the function $M_A : \Omega \longrightarrow \bar{\mathbb{R}}_+$ such that $M_A(\omega) = M(\omega, A)$ for all $\omega \in \Omega$ is a random variable on $(\Omega, \mathcal{A})$.*

A random measure M is said to be (purely) atomic as soon as the measure $\bar{M}_\omega$ is (purely) atomic for almost every $\omega \in \Omega$.

EXAMPLE 2.13.– Let $(\mathbf{X}_n)_{n\geq 1}$ be a sequence of random variables from $(\Omega, \mathcal{A})$ to $\left(\mathbb{R}_+^d, \mathcal{B}\left(\mathbb{R}_+^d\right)\right)$ and let

$$M = \sum_{n=1}^{+\infty} \delta_{\mathbf{X}_n}. \qquad [2.2]$$

For all $A \in \mathcal{B}\left(\mathbb{R}_+^d\right)$ and all $\omega \in \Omega$:

$$M_A(\omega) = M(\omega, A) = \sum_{n=1}^{+\infty} \mathbf{1}_A(\mathbf{X}_n(\omega)), \qquad [2.3]$$

$$\bar{M}_\omega(A) = M(\omega, A) = \sum_{n=1}^{+\infty} \delta_{\mathbf{X}_n(\omega)}(A) \qquad [2.4]$$

and $M(\omega, A)$ is the number of points $\mathbf{X}_n(\omega)$ that belong to the Borel set A:

$$M(\omega, A) = \#\{n \geq 1 : \mathbf{X}_n(\omega) \in A\}.$$

Based on [2.3], the function $M_A(\cdot) = M(\cdot, A)$ is measurable for each $A \in \mathcal{B}(\mathbb{R}_+^d)$. Based on [2.4], the function $\bar{M}_\omega(\cdot) = M(\omega, \cdot)$ is a measure of $(\mathbb{R}_+^d, \mathcal{B}(\mathbb{R}_+^d))$. As a result, the function M defined by [2.2] is a random measure. It is an atomic random measure as soon as for almost each $\omega \in \Omega$ and each compact set B, the points $\mathbf{X}_n(\omega)$ that belong to B are finitely many. It is a purely atomic measure if furthermore, the $\mathbf{X}_n$'s are almost surely distinct.

REMARK 2.14.– A random measure seen as a random variable.

Setting $\mathcal{M}$ to be the space of measures on $(\mathbb{R}_+^d, \mathcal{B}(\mathbb{R}_+^d))$ and based on the first point of the definition, a random measure M can be seen as a function from Ω to $\mathcal{M}$, with $M(\omega) = \bar{M}_\omega$. Assume $\mathcal{M}$ to be equipped with the σ-algebra $\mathcal{A}_\mathcal{M}$ generated by the functions $F_A : \mathcal{M} \to (\mathbb{R}_+, \mathcal{B}(\mathbb{R}_+))$, with $F_A(\mu) = \mu(A)$ for all $A \in \mathcal{B}(\mathbb{R}_+^d)$ and all $\mu \in \mathcal{M}$. Recall that a function $G : (\Omega, \mathcal{A}) \to (\mathcal{M}, \mathcal{A}_\mathcal{M})$ is measurable if and only if $F_A \circ G$ is measurable from $(\Omega, \mathcal{A})$ to $(\mathbb{R}_+, \mathcal{B}(\mathbb{R}_+))$ for all $A \in \mathcal{B}(\mathbb{R}_+^d)$. The second point of the definition now means that the function $M_A : \omega \to M(\omega, A) = (F_A \circ M)(\omega)$ is measurable from $(\Omega, \mathcal{A})$ to $(\mathbb{R}_+, \mathcal{B}(\mathbb{R}_+))$ for all $A \in \mathcal{B}(\mathbb{R}_+^d)$, and consequently that $M : (\Omega, \mathcal{A}) \to (\mathcal{M}, \mathcal{A}_\mathcal{M})$ is measurable. A random measure M can hence be seen as a random variable from $(\Omega, \mathcal{A})$ to $(\mathcal{M}, \mathcal{A}_\mathcal{M})$, hence the name "random measure".

REMARK 2.15.– Characterization of the distribution of a random measure.

Based on the previous remark, the distribution of M is fully determined by the distribution of the collection of random variables M_A where A is any Borel set, or equivalently, by the distributions of all random vectors of finite length $(M_{A_1}, \cdots, M_{A_n})$, where $A_1, \cdots, A_n \in \mathcal{B}(\mathbb{R}_+^d)$ and $n \in \mathbb{N}^*$. Note that any random vector $(M_{A_1}, \cdots, M_{A_n})$ may always be expressed with respect to a similar random vector with pairwise disjoint Borel sets. (For instance, for $n = 2$, take $B_1 = A_1 \cap A_2$, $B_2 = A_1 \cap \bar{A}_2$, $B_3 = \bar{A}_1 \cap A_2$; then $(M_{A_1}, M_{A_2}) = (M_{B_1} + M_{B_2}, M_{B_1} + M_{B_3})$.). As a result, the distribution of M is fully determined by the distribution of the random vectors $(M_{A_1}, \cdots, M_{A_n})$, where $A_1, \cdots, A_n$ are now restricted to be pairwise disjoint Borel sets. Moreover, let $\mathcal{C}$ be a generating set of $\mathcal{B}(\mathbb{R}_+^d)$, stable by finite intersection and such that there exists a sequence of locally compact Borel sets $(C_n)_{n \in \mathbb{N}}$ in $\mathcal{C}$ that increases toward $\mathbb{R}_+^d$ and such that $\bar{M}_\omega(C_n) < +\infty$ for all $n \in \mathbb{N}^*$ and almost all $\omega \in \Omega$. We may then prove that it is enough to check the measurability property of the functions M_A only for A in $\mathcal{C}$, for the function $M : \Omega \to \mathcal{M}$ to be a random measure. Also, the previous characterizations of the distribution of M remain valid considering only Borel sets belonging to $\mathcal{C}$ (see [NEV 77, Prop. I.2]).

We now come to the definition of Poisson random measures, where we limit ourselves to the case of purely atomic random measures.

DEFINITION 2.16.– *Let $M : \Omega \times \mathcal{B}\left(\mathbb{R}_+^d\right) \longrightarrow \bar{\mathbb{R}}_+$ be a purely atomic (or counting) random measure and let ν be a σ-finite measure on $\left(\mathbb{R}_+^d, \mathcal{B}\left(\mathbb{R}_+^d\right)\right)$. The function M is a Poisson random measure with intensity (or mean) measure ν if:*

1) *for all $n \geq 2$ and all pairwise disjoint Borel sets $A_1, \cdots, A_n \in \mathcal{B}\left(\mathbb{R}_+^d\right)$, the random variables $M_{A_1}, \cdots, M_{A_n}$ are independent;*

2) *for each $A \in \mathcal{B}\left(\mathbb{R}_+^d\right)$, the random variable M_A is Poisson distributed with mean $\nu(A)$.*

Based on remark 2.15, the distribution of any random measure M is fully determined by the distributions of the random vectors $(M_{A_1}, \cdots, M_{A_n})$, where $A_1, \cdots, A_n$ are pairwise disjoint Borel sets. In case of a Poisson random measure, the r.v.s M_{A_i}, $1 \leq i \leq n$ are independent and the distribution of $(M_{A_1}, \cdots, M_{A_n})$ is simply the product of the distributions of the M_{A_i}'s, which are Poisson distributed $\mathcal{P}(\nu(A_i))$. As a result, the distribution of a Poisson random measure is fully determined by its intensity measure ν.

EXAMPLE 2.17.– Let $(\mathbf{X}_n)_{n \geq 1}$ be a sequence of i.i.d. r.v.s on $\mathbb{R}_+^d$ with common distribution μ, which is assumed to be σ-finite and absolutely continuous with respect to Lebesgue measure. This ensures the $\mathbf{X}_n$'s to be almost surely distinct. Let K be an independent Poisson distributed r.v. $\mathcal{P}(\lambda)$, with $\lambda > 0$. We set:

$$M = \sum_{n=1}^{K} \delta_{\mathbf{X}_n},$$

which means that

$$M(\omega, A) = \sum_{n=1}^{K(\omega)} \delta_{\mathbf{X}_n(\omega)}(A) = \sum_{n=1}^{K(\omega)} \mathbf{1}_A(\mathbf{X}_n(\omega))$$

for all $A \in \mathcal{B}\left(\mathbb{R}_+^d\right)$ and all $\omega \in \Omega$. Using the same arguments as for example 2.13, it is easy to see that M is a random measure. As the $\mathbf{X}_n$'s are almost surely distinct and as $K(\omega) < +\infty$ almost surely, the random measure M is purely atomic. Let us now show that M is Poisson. Let $\{A_1, \cdots, A_n\}$ be a Borel partition of $\mathbb{R}_+^d$ and note that $\sum_{j=1}^n M_{A_j} = K$. Let

$(i_1, \cdots, i_n) \in \mathbb{N}^n$ and set $k = \sum_{j=1}^n i_j$. Using the independence assumptions and a multinomial model for the third equality, we get:

$$\begin{aligned}
&\mathbb{P}(M_{A_1} = i_1, \cdots, M_{A_n} = i_n) \\
&= \mathbb{P}(M_{A_1} = i_1, \cdots, M_{A_n} = i_n | K = k)\, \mathbb{P}(K = k) \\
&= \mathbb{P}\left(\sum_{n=1}^{k} \mathbf{1}_{A_1}(\mathbf{X}_n) = i_1, \cdots, \sum_{n=1}^{k} \mathbf{1}_{A_n}(\mathbf{X}_n) = i_n\right) \mathbb{P}(K = k) \\
&= \frac{k!}{i_1! \cdots i_n!} \mu(A_1)^{i_1} \times \cdots \times \mu(A_n)^{i_n} \times e^{-\lambda} \frac{\lambda^k}{k!} \\
&= \prod_{j=1}^{n} \frac{(\lambda\, \mu(A_j))^{i_j}}{i_j!} e^{-\lambda\, \mu(A_j)}
\end{aligned}$$

using that $\sum_{j=1}^n \mu(A_j) = 1$ for the last line. This proves that the r.v.s M_{A_1}, $\cdots$, M_{A_n} are independent and that M_{A_j} is Poisson distributed $\mathcal{P}(\lambda\, \mu(A_j))$ for all $1 \le j \le n$. As every set $\{A_1, \cdots, A_m\}$ of pairwise disjoint Borel sets can be completed into a partition of $\mathbb{R}_+^d$, the result is hence true for any finite set of pairwise disjoint Borel sets. As a result, M is a Poisson random measure with intensity measure $\nu(\mathrm{d}\mathbf{u}) = \lambda\, \mu(\mathrm{d}\mathbf{u})$.

Example 2.17 is the base of a constructive proof of existence of a Poisson random measure with a given σ-finite intensity measure, see [ÇIN 11] or [SAT 99] for more details. Another way to explicitly construct Poisson random measures will be provided by corollary 2.22 later on.

We now go on with another result that makes the connection clear between Poisson random measures and standard Poisson processes on $\mathbb{R}_+$.

PROPOSITION 2.18.– Let $\Lambda : \mathbb{R}_+ \to \mathbb{R}_+$ be a non-decreasing and continuous function such that $\Lambda(0) = 0$ and let $\lambda(du)$ be the single diffuse and σ-finite measure on $(\mathbb{R}_+, \mathcal{B}(\mathbb{R}_+))$ such that $\Lambda(t) = \lambda([0,t])$, for all $t \ge 0$. Let M be a purely atomic random measure and let $N_t = M_{[0,t]}$, for all $t \ge 0$. Then, M is a Poisson random measure with intensity measure $\lambda(du)$ if and only if $(N_t)_{t \ge 0}$ is a non-homogeneous Poisson process with mean function $\Lambda(t)$. In case $\Lambda(t) = \lambda t$ with $\lambda > 0$, we get that M is a Poisson random measure with intensity measure $\lambda\, du$ if and only if $(N_t)_{t \ge 0}$ is a homogeneous Poisson process with rate λ.

PROOF.– Note first that based on the continuity of $\Lambda(t)$, we have $\lambda(\{t\}) = \Lambda(t) - \Lambda(t^-) = 0$ for all $t \ge 0$, so that $\lambda(du)$ is necessary diffuse. Also,

$\lambda([0,n]) = \Lambda(n) < +\infty$ for all $n \in \mathbb{N}^*$ with $\cup_{n\in\mathbb{N}^*}[0,n] = \mathbb{R}_+$ and the measure $\lambda(\mathrm{d}u)$ is also σ-finite. Now, assume that M is a Poisson measure with intensity measure $\lambda(\mathrm{d}u)$ and let us show that $(N_t)_{t\geq 0}$ is a Poisson process with mean rate function $\Lambda(t)$. First, as M is a purely atomic random measure, $(N_t)_{t\geq 0}$ is a counting process (namely with jumps of size 1). Also, based on the definition of a Poisson random measure and taking $0 \leq t_1 < \cdots < t_{n+1}$, the events $A_i = (t_i, t_{i+1}]$ for $i = 1, \cdots, n$ are pairwise disjoint so that the r.v.s $M_{(t_i,t_{i+1}]} = N_{t_{i+1}} - N_{t_i}$ for $i = 1, \cdots, n$ are independent. Also, the r.v. $N_{t_{i+1}} - N_{t_i}$ is Poisson distributed with mean

$$\lambda((t_i, t_{i+1}]) = \Lambda(t_{i+1}) - \Lambda(t_i).$$

The process $(N_t)_{t\geq 0}$ is consequently a non-homogeneous Poisson process with mean rate function $\Lambda(t)$. Conversely, assume that $(N_t)_{t\geq 0}$ is a Poisson process with mean rate function $\Lambda(t)$ and let $(T_n)_{n\geq 1}$ be the points of $(N_t)_{t\geq 0}$. For each $\omega \in \Omega$, we have

$$\begin{aligned} \bar{M}_\omega([0,t]) &= M_{[0,t]}(\omega) = N_t(\omega) \\ &= \sum_{n=1}^{+\infty} \delta_{T_n(\omega)}([0,t]) < +\infty, \text{ for all } t \geq 0. \end{aligned} \qquad [2.5]$$

The two measures $\bar{M}_\omega$ and $\sum_{n=1}^{+\infty} \delta_{T_n(\omega)}$ on $(\mathbb{R}_+, \mathcal{B}(\mathbb{R}_+))$ hence agree and are finite on $\mathcal{C} = \{[0,t], t \in \mathbb{R}_+\}$. As $\mathcal{C}$ is a generating set of $\mathcal{B}(\mathbb{R}_+)$, stable by finite intersection, and such that there exists a sequence of sets in $\mathcal{C}$ that increases toward $\mathbb{R}_+$, we conclude that the two measures agree on the whole σ-algebra $\mathcal{B}(\mathbb{R}_+)$ (see, e.g. [BIL 95, theorem 10.3]):

$$\bar{M}_\omega = \sum_{n=1}^{+\infty} \delta_{T_n(\omega)}, \text{ for all } \omega \in \Omega.$$

Equivalently, we have:

$$M = \sum_{n=1}^{+\infty} \delta_{T_n}.$$

The point now is to check that M is Poisson. Based on remark 2.15 and on the properties of the set $\mathcal{C} = \{[0,t], t \in \mathbb{R}_+\}$ previously seen, the distribution of M is fully determined by the distributions of the random vectors $\left(M_{[0,t_1]}, \cdots, M_{[0,t_n]}\right)$, or equivalently, by the distributions of the random vectors $\left(M_{[0,t_1]}, M_{(t_1,t_2]}, \cdots, M_{(t_{n-1},t_n]}\right)$, where $0 \leq t_1 < \cdots$

$< t_n$. As $(N_t)_{t \geq 0}$ is a Poisson process, the distributions of these random vectors are already known to be in accordance with the Poisson property and we conclude that M is Poisson with measure intensity $\lambda(du)$. □

The following proposition provides a way to identify independent Poisson random measures.

PROPOSITION 2.19.– Let N be a Poisson random measure with intensity measure $\mu(d\mathbf{u})$ on $\mathbb{R}_+^d$. Let B_1, ..., B_n be disjoint Borel sets of $\mathbb{R}^d$ and let N^{B_i} be the restriction of N to B_i, with

$$N^{B_i}(\omega, A) = N(\omega, A \cap B_i)$$

for all $\omega \in \Omega$, all $A \in \mathcal{B}(\mathbb{R}_+^d)$ and all $1 \leq i \leq n$. Then, N^{B_i} is a Poisson random measure with intensity measure $\mu^{B_i}(d\mathbf{u})$, where

$$\mu^{B_i}(A) = \mu(A \cap B_i), \text{ for all } A \in \mathcal{B}(\mathbb{R}_+^d) \text{ and all } 1 \leq i \leq n.$$

Also, the Poisson random measures N^{B_1}, ..., N^{B_n} are independent.

PROOF.– Let $1 \leq i \leq n$ and let us first check that N^{B_i} is a Poisson random measure with intensity measure $\mu^{B_i}(d\mathbf{u})$. With that aim, let A_1, ..., A_m be disjoint Borel sets of $\mathbb{R}^d$. Then:

$$N^{B_i}(\omega, A_j) = N(\omega, A_j \cap B_i) = N_{A_j \cap B_i}(\omega), \text{ for all } 1 \leq j \leq m.$$

The Borel sets $A_1 \cap B_i$, $A_2 \cap B_i$, ..., $A_m \cap B_i$ are pairwise disjoint so that the r.v.s $N_{A_j \cap B_i} = N_{A_j}^{B_i}$ for $1 \leq j \leq m$ are independent (based on the Poisson property of N). Also, for each j, the r.v. $N_{A_j \cap B_i} = N_{A_j}^{B_i}$ is Poisson distributed with mean $\mu(A_j \cap B_i) = \mu^{B_i}(A_j)$. This proves that N^{B_i} is a Poisson random measure with intensity measure $\mu^{B_i}(d\mathbf{u})$. To prove the independence of the Poisson random measures N^{B_1}, ..., N^{B_n}, it is sufficient to prove that the r.v.s $N^{B_1}(C_1)$, ..., $N^{B_n}(C_n)$ are independent, for all Borel sets C_1, ..., C_m. As $N^{B_i}(C_i) = N(C_i \cap B_i)$ and as the sets $C_i \cap B_i$, $1 \leq i \leq n$ are disjoint, the property is clear, based on the Poisson property of N. □

Before going on with the study of random measures, we introduce a new notation and, for any random measure M and any measurable function f: $(\mathbb{R}_+^d, \mathcal{B}(\mathbb{R}_+^d)) \longrightarrow (\mathbb{R}_+, \mathcal{B}(\mathbb{R}_+))$, we set Mf to be the r.v. such that

$$(Mf)(\omega) = \int_{\mathbb{R}_+^d} f(\mathbf{u}) M(\omega, d\mathbf{u}), \text{ for all } \omega \in \Omega.$$

Note that

$$M(\alpha f + \beta g) = \alpha\, Mf + \beta\, Mg \qquad [2.6]$$

for all $\alpha, \beta \geq 0$ and all measurable non-negative functions f and g.

In case of a purely atomic random measure $M = \sum_{n=1}^{+\infty} \delta_{\mathbf{X}_n}$, we get:

$$\begin{aligned}(Mf)(\omega) &= \int_{\mathbb{R}_+^d} f(\mathbf{u}) \sum_{n=1}^{+\infty} \delta_{\mathbf{X}_n(\omega)}(d\mathbf{u}) \\ &= \sum_{n=1}^{+\infty} \int_{\mathbb{R}_+^d} f(\mathbf{u})\, \delta_{\mathbf{X}_n(\omega)}(d\mathbf{u}) \\ &= \sum_{n=1}^{+\infty} f(\mathbf{X}_n(\omega)),\end{aligned}$$

for all $\omega \in \Omega$, based on Fubini's theorem.

We now define the *Laplace functional* of a random measure, which provides an easy characterization of its distribution.

THEOREM 2.20.– The probability distribution of a random measure M is characterized by its Laplace functional $\mathcal{L}_M$, where

$$\mathcal{L}_M(f) = \mathbb{E}\left(e^{-Mf}\right)$$

for all non-negative measurable function f.

PROOF.– Based on remark 2.15, the distribution of a random measure M is characterized by the distributions of the random vectors $(M_{A_1}, \cdots, M_{A_n})$, where $n \in \mathbb{N}^*$ and $A_1, \cdots, A_n$ are Borel sets. Let us remember that the distribution of any non-negative random vector $(X_1, \cdots, X_n)$ is characterized by its multivariate Laplace transform

$$\mathcal{L}_{(X_1,\cdots,X_n)}(s_1, \cdots, s_n) = \mathbb{E}\left(e^{-\sum_{i=1}^n s_i\, X_i}\right), \text{ for all } (s_1, \cdots, s_n) \in \mathbb{R}_+^n.$$

The distribution of M is hence characterized by the collection of non-negative real numbers of the form

$$\mathbb{E}\left(e^{-\sum_{i=1}^n s_i\, M_{A_i}}\right)$$

where $n \in \mathbb{N}^*$, $A_1, \cdots, A_n$ are Borel sets and $(s_1, \cdots, s_n) \in \mathbb{R}_+^n$. Setting

$$f = \sum_{i=1}^{n} s_i \, \mathbf{1}_{A_i} \tag{2.7}$$

and noting that

$$Mf = \sum_{i=1}^{n} s_i \, M_{A_i}$$

due to [2.6], the distribution of M is hence characterized by the collection of real numbers $\mathbb{E}\left(e^{-Mf}\right)$, where f is any non-negative measurable function of the shape [2.7]. It is hence characterized by the same collection when the set of test functions f is enlarged to be the set of all non-negative measurable functions. □

A Poisson random measure is characterized by a specific shape for its Laplace functional, as provided by the following theorem.

THEOREM 2.21.– Let M be a purely atomic random measure on $\Omega \times \mathcal{B}\left(\mathbb{R}_+^d\right)$ and let ν be a σ-finite measure on $\left(\mathbb{R}_+^d, \mathcal{B}\left(\mathbb{R}_+^d\right)\right)$. The mapping M is a Poisson random measure with intensity measure ν if and only if

$$\mathbb{E}\left(e^{-Mf}\right) = e^{-\nu\left(1-e^{-f}\right)} \tag{2.8}$$

for all measurable function $f : \left(\mathbb{R}_+^d, \mathcal{B}\left(\mathbb{R}_+^d\right)\right) \longrightarrow \left(\mathbb{R}_+, \mathcal{B}\left(\mathbb{R}_+\right)\right)$, where

$$\nu\left(1 - e^{-f}\right) = \int_{\mathbb{R}_+^d} \left(1 - e^{-f(\mathbf{u})}\right) \nu\left(d\mathbf{u}\right).$$

PROOF.– Let B_1, ..., B_n be pairwise disjoint Borel sets and let $f = \sum_{i=1}^{n} x_i \mathbf{1}_{B_i}$, where $x_i \in \mathbb{R}_+$, for all $1 \leq i \leq n$. Based on [2.6], we have:

$$Nf = \sum_{i=1}^{n} x_i N_{B_i}$$

and

$$\mathbb{E}\left(e^{-Nf}\right) = \mathbb{E}\left(e^{-\sum_{i=1}^{n} x_i N_{B_i}}\right). \tag{2.9}$$

Besides

$$\nu\left(1-e^{-f}\right)=\int_{\mathbb{R}^d}(1-e^{-f(\mathbf{y})})\nu(d\mathbf{y})=\sum_{j=1}^{n}\int_{B_j}(1-e^{-f(\mathbf{y})})\nu(d\mathbf{y})$$

because $1-e^{-f(\mathbf{y})}=0$ when $\mathbf{y}\notin\cup_{j=1}^{n}B_j$. For $\mathbf{y}\in B_j$, we have $1-e^{-f(\mathbf{y})}=1-e^{-x_j}$ so that

$$\nu\left(1-e^{-f}\right)=\sum_{j=1}^{n}\int_{B_j}(1-e^{-x_j})\nu(d\mathbf{y})=\sum_{j=1}^{n}\nu(B_j)(1-e^{-x_j})$$

and

$$e^{-\nu\left(1-e^{-f}\right)}=\prod_{j=1}^{n}e^{-\nu(B_j)(1-e^{-x_j})}. \qquad [2.10]$$

Now, assume that [2.8] is true. Then, based on [2.9] and [2.10]:

$$\mathbb{E}\left(e^{-\sum_{i=1}^{n}x_iN_{B_i}}\right)=\prod_{j=1}^{n}e^{-\nu(B_j)(1-e^{-x_j})}.$$

This proves that the r.v.s N_{B_i}, $i=1,\cdots,n$, are independent and that N_{B_i} is Poisson distributed $\mathcal{P}\left(\nu\left(B_i\right)\right)$. (Remember the Laplace transform of a Poisson r.v. provided in [2.1].) Hence, N is a Poisson random measure with intensity measure ν. Conversely, assume that N is a Poisson random measure with intensity measure ν. Then, the r.v.s N_{B_i}, $i=1,\cdots,n$, are independent and each N_{B_i} is Poisson distributed $\mathcal{P}\left(\nu\left(B_i\right)\right)$. This provides:

$$\mathbb{E}\left(e^{-\sum_{i=1}^{n}x_iN_{B_i}}\right)=\prod_{i=1}^{n}\mathbb{E}\left(e^{-x_iN_{B_i}}\right)=\prod_{i=1}^{n}e^{-\nu(B_i)\left(1-e^{-x_i}\right)}$$

due to [2.1]. Based on [2.9] and [2.10] again, we get:

$$\mathbb{E}\left(e^{-Nf}\right)=e^{-\nu\left(1-e^{-f}\right)}$$

for all piecewise constant measurable non-negative function $f=\sum_{i=1}^{n}x_i\mathbf{1}_{B_i}$. We easily derive the result for each measurable non-negative function f by Lebesgue's dominated convergence theorem using the fact that f is the limit of a non-decreasing sequence of piecewise constant measurable non-negative functions. □

COROLLARY 2.22.– Let $N = \sum_{n\geq 1} \delta_{\mathbf{X}_n}$ be a Poisson random measure with (σ-finite) intensity measure $\mu(\mathrm{d}\mathbf{x})$ on $\mathbb{R}_+^d$ and let $(\mathbf{Y}_n)_{n\in\mathbb{N}^*}$ be a sequence of i.i.d. r.v.s on $\mathbb{R}_+^m$ with common distribution $\nu(d\mathbf{y})$ and independent of N (or of the $\mathbf{X}_n$'s, equivalently). Then, $M = \sum_{n\geq 1} \delta_{(\mathbf{X}_n, \mathbf{Y}_n)}$ is a Poisson random measure on $\mathbb{R}_+^d \times \mathbb{R}_+^m$ with measure intensity $\mu \otimes \nu$, where we recall that the product measure $\mu \otimes \nu$ is the single measure on $\mathcal{B}(\mathbb{R}_+^d \times \mathbb{R}_+^m)$ such that:

$$(\mu \otimes \nu)(B_1 \times B_2) = \mu(B_1)\,\nu(B_2)$$

for all $B_1 \in \mathcal{B}(\mathbb{R}_+^d)$ and all $B_2 \in \mathcal{B}(\mathbb{R}_+^m)$.

PROOF.– M clearly is a purely atomic measure as N. Also, $\mu \otimes \nu$ is σ-finite because μ is σ-finite and ν is a probability measure. The point is hence to prove that M is Poisson. Let f be a non-negative Borel function. Conditioning by the σ-algebra $\sigma(\mathbf{X}_m, m \geq 1)$ generated by $\{\mathbf{X}_m, m \geq 1\}$ and based on the independence assumptions, we have:

$$\begin{aligned}
\mathbb{E}\left(e^{-Mf}\right) &= \mathbb{E}\left(e^{-\sum_{n\geq 1} f(\mathbf{X}_n,\mathbf{Y}_n)}\right) \\
&= \mathbb{E}\left[\mathbb{E}\left(e^{-\sum_{n\geq 1} f(\mathbf{X}_n,\mathbf{Y}_n)} | \sigma(\mathbf{X}_m, m \geq 1)\right)\right] \\
&= \mathbb{E}\left[\prod_{n\geq 1} \mathbb{E}\left(e^{-f(\mathbf{X}_n,\mathbf{Y}_n)} | \mathbf{X}_n\right)\right] \\
&= \mathbb{E}\left(\prod_{n\geq 1} g(\mathbf{X}_n)\right)
\end{aligned} \qquad [2.11]$$

with

$$g(\mathbf{x}) = \mathbb{E}\left(e^{-f(\mathbf{x},\mathbf{Y}_n)}\right) = \int_{\mathbb{R}_+^m} e^{-f(\mathbf{x},\mathbf{y})}\nu(\mathrm{d}\mathbf{y}), \text{ for all } \mathbf{x} \in \mathbb{R}_+^d.$$

Based on [2.11], we have:

$$\mathbb{E}\left(e^{-Mf}\right) = \mathbb{E}\left(e^{\sum_{n\geq 1} \log(g(\mathbf{X}_n))}\right) = \mathbb{E}\left(e^{-Nh}\right)$$

with

$$h(\mathbf{x}) = -\log(g(\mathbf{x})), \text{ for all } \mathbf{x} \in \mathbb{R}_+^d.$$

As N is a Poisson random measure with intensity measure μ, we get from theorem 2.21 that

$$\mathbb{E}\left(e^{-Mf}\right) = e^{-\int_{\mathbb{R}_+^d}\left(1-e^{-h(\mathbf{x})}\right)\mu(\mathrm{d}\mathbf{x})} = e^{-\int_{\mathbb{R}_+^d}(1-g(\mathbf{x}))\mu(\mathrm{d}\mathbf{x})} \qquad [2.12]$$

where

$$\begin{aligned}\int_{\mathbb{R}_+^d}(1-g(\mathbf{x}))\,\mu(\mathrm{d}\mathbf{x}) &= \int_{\mathbb{R}_+^d}\left(1-\int_{\mathbb{R}_+^m} e^{-f(\mathbf{x},\mathbf{y})}\nu(\mathrm{d}\mathbf{y})\right)\mu(\mathrm{d}\mathbf{x})\\ &= \int_{\mathbb{R}_+^d}\left(\int_{\mathbb{R}_+^m}\left(1-e^{-f(\mathbf{x},\mathbf{y})}\right)\nu(\mathrm{d}\mathbf{y})\right)\mu(\mathrm{d}\mathbf{x})\\ &= \int_{\mathbb{R}_+^d\times\mathbb{R}_+^m}\left(1-e^{-f(\mathbf{x},\mathbf{y})}\right)(\mu\otimes\nu)(\mathrm{d}\mathbf{x},\mathrm{d}\mathbf{y})\\ &= (\mu\otimes\nu)\left(1-e^{-f}\right)\end{aligned}$$

using that $\nu\left(\mathbb{R}_+^m\right) = 1$ for the second equality. Substituting this last expression into [2.12] provides the result. □

Let $\lambda(\mathrm{d}u)$ be a diffuse and σ-finite measure on $(\mathbb{R}_+, \mathcal{B}(\mathbb{R}_+))$ and let $\nu(\mathrm{d}\mathbf{v})$ be a bounded measure on $\left(\mathbb{R}_+^d, \mathcal{B}\left(\mathbb{R}_+^d\right)\right)$, with $0 < \nu\left(\mathbb{R}_+^d\right) = C < +\infty$. The previous result along with proposition 2.18 provides an easy way to construct the points of a Poisson random measure on $\mathbb{R}_+^{d+1}$ with intensity measure of the form

$$(\lambda\otimes\nu)(\mathrm{d}u, \mathrm{d}\mathbf{v}) = \lambda(\mathrm{d}u)\ \nu(\mathrm{d}\mathbf{v}).$$

Indeed, let us normalize $\nu(\mathrm{d}\mathbf{v})$ into the probability measure $\mu(\mathrm{d}\mathbf{v}) = \frac{\nu(\mathrm{d}\mathbf{v})}{C}$ and let us set $\alpha(\mathrm{d}u) = C\,\lambda(\mathrm{d}u)$ so that

$$(\lambda\otimes\nu)(\mathrm{d}u, \mathrm{d}\mathbf{v}) = (\alpha\otimes\mu)(\mathrm{d}u, \mathrm{d}\mathbf{v}).$$

Let $A(t) = \alpha([0,t])$ for all $t \geq 0$ and let $(N_t)_{t\geq 0}$ be a Poisson process with mean function $A(t)$. We set $(U_n)_{n\geq 1}$ to be the points of $(N_t)_{t\geq 0}$ and $N = \sum_{n\geq 1}\delta_{U_n}$.

Based on proposition 2.18 (and on its proof), N is a Poisson random measure with intensity measure $\alpha(\mathrm{d}u)$. Considering a sequence $(\mathbf{V}_n)_{n\geq 1}$ of i.i.d. r.v.s on $\left(\mathbb{R}_+^d, \mathcal{B}\left(\mathbb{R}_+^d\right)\right)$ with distribution $\mu(\mathrm{d}\mathbf{v})$ and independent of N, the random measure $M = \sum_{n\geq 1} \delta_{(U_n, \mathbf{V}_n)}$ is Poisson with intensity measure $(\alpha \otimes \mu) = (\lambda \otimes \nu)$ based on corollary 2.22. The points of M that belong to $[0, T] \times \mathbb{R}_+^d$ are now easy to simulate based on algorithm 2.10. This provides the following algorithm (assuming $A(t)$ to be one-to-one).

ALGORITHM 2.23.– Simulation of an observation $(u_i, \mathbf{v}_i)_{1\leq i\leq n}$ of the points of the Poisson random measure M on $[0, T] \times \mathbb{R}_+^d$:

– generate an observation n of N_T according to $\mathcal{P}(A(T))$;

– generate n i.i.d. observations $x_1, \cdots, x_n$ according to $\mathcal{U}([0, 1])$ (independently of n);

– set $u_i = A^{-1}(A(t)\, x_i)$ for all $1 \leq i \leq n$;

– generate n i.i.d. observations $\mathbf{v}_1, \cdots, \mathbf{v}_n$ with distribution $\mu(d\mathbf{v})$ (independently of n and of the u_i's).

If needed, we can also add the sorting step of the points $x_1, \cdots, x_n$ of the Poisson process $(N_t)_{t\geq 0}$ as in algorithm 2.10.

We are now ready to define and study gamma processes.

2.3. Definition and basic properties of a gamma process

DEFINITION 2.24.– *Let $A : \mathbb{R}^+ \longrightarrow \mathbb{R}^+$ be a non-decreasing and right-continuous function such that $A(0) = 0$ and let $b > 0$. Let $(X_t)_{t\geq 0}$ be a right-continuous stochastic process with left-side limits (càdlag process). Then, $(X_t)_{t\geq 0}$ is called a (non-homogeneous) gamma process with shape function $A(t)$ and rate parameter b ($\mathcal{G}(A(t), b)$ for short) if:*

– $X_0 = 0$ almost surely;

– $(X_t)_{t\geq 0}$ is a process with independent increments: for all $n \in \mathbb{N}^*$, all $(t_1, \cdots, t_{n+1}) \in \mathbb{R}^{n+1}$ such that $0 \leq t_1 < t_2 < \cdots < t_{n+1}$, the increments $X_{t_{i+1}} - X_{t_i}$ with $1 \leq i \leq n$ are independent;

– for all $0 < s < t$, the r.v. $X_t - X_s$ is gamma distributed $\mathcal{G}(A(t) - A(s), b)$.

In case $A(t) = at$ for all $t \geq 0$ with $a > 0$, the increment $X_t - X_s$ is gamma distributed $\mathcal{G}(a(t - s), b)$ and the process is called a homogeneous gamma process with parameters (a, b).

Let us first remark that the third point of the definition implies that, for all $0 < s < t$, the r.v. $X_t - X_s$ is non-negative. As a result, a gamma process is a *non-decreasing* process.

Also, considering $0 \leq s < t < u$, the gamma preservation property from corollary 2.3 implies that the random variable

$$X_u - X_s = (X_u - X_t) + (X_t - X_s)$$

is gamma distributed with shape parameter

$$(A(u) - A(t)) + (A(t) - A(s)) = A(u) - A(s),$$

based on the independence between $X_t - X_s$ and $X_u - X_t$ and on the fact that their respective distributions are $\mathcal{G}(A(t) - A(s), b)$ and $\mathcal{G}(A(u) - A(t), b)$. The definition of a gamma process hence is coherent with this gamma preservation property.

Finally, as will be seen in the following, a gamma process is stochastically continuous as soon as the shape function $A(t)$ is continuous. In that case, a gamma process is an additive process (see definition I.1 in the Introduction). The distribution of X_t is hence known to be infinitely divisible [SAT 99]. In the specific case of a *homogeneous* gamma process (with $A(t) = at$), the distribution of $X_t - X_s$ only depends on $t - s$ for all $0 \leq s \leq t$ and the increments are stationary. A homogeneous gamma process is hence a Lévy process. The infinitely divisibility property is here clear, writing

$$X_t = \sum_{i=1}^{n} (X_{\frac{i}{n}t} - X_{\frac{i-1}{n}t})$$

and noting that the $X_{\frac{i}{n}t} - X_{\frac{i-1}{n}t}$'s are i.i.d. gamma distributed $\mathcal{G}(a\frac{t}{n}, b)$.

Based on corollary 2.5, we get directly the following scaling result, which often allows us to concentrate on the special case $b = 1$, with no restriction.

PROPOSITION 2.25.–

1) Let $(X_t)_{t \geq 0}$ be a gamma process $\mathcal{G}(A(t), b)$ and let $c > 0$. Then, $(c\,X_t)_{t \geq 0}$ is a gamma process $\mathcal{G}\left(A(t), \frac{b}{c}\right)$.

2) Let $b > 0$ and $Y_t = bX_t$, for all $t \geq 0$. Then, the process $(X_t)_{t\geq 0}$ is a gamma process $\mathcal{G}(A(t), b)$ if and only if $(Y_t)_{t\geq 0}$ is a gamma process $\mathcal{G}(A(t), 1)$.

Based on proposition 2.8, we directly get the following comparison result, which may be, e.g. useful for the theoretical study of preventive maintenance policies.

PROPOSITION 2.26.– Let $(X_t)_{t\geq 0}$ be a gamma process $\mathcal{G}(A(t), b)$. Then, if $0 \leq t_1 \leq t_2$, the r.v. X_{t_1} is smaller than X_{t_2} in the sense of the likelihood ratio order and, consequently, in the sense of the (reversed) hazard rate and usual stochastic orders too.

The remaining of this section can be skipped in the first reading.

In the definition, we have assumed that $A : \mathbb{R}^+ \longrightarrow \mathbb{R}^+$ is a non-decreasing and right-continuous function such that $A(0) = 0$. The function A is hence continuous on $\mathbb{R}^+$, except on a set $\mathcal{I} \subset \mathbb{R}^+$ which is at most countable. Following [ÇIN 80], we can set A^c and A^d to be the continuous and discontinuous parts of A, with

$$A^d(t) = \sum_{s\in\mathcal{I}, s\leq t} \Delta A(s),$$

$$A^c(t) = A(t) - A^d(t),$$

where $\Delta A(s) = A(s) - A(s^-)$.

Introducing $(X_t^c)_{t\geq 0}$ and $\left(X_t^d\right)_{t\geq 0}$ as two independent gamma processes with respective parameters $(A^c(t), b)$ and $\left(A^d(t), b\right)$ and using again corollary 2.3, the random variable X_t with distribution $\mathcal{G}\left(A^c(t) + A^d(t), b\right)$ is identically distributed as the sum $X_t^c + X_t^d$:

$$X_t \overset{d}{=} X_t^c + X_t^d. \qquad [2.13]$$

Let us now look at the fixed discontinuities of $\left(X_t^d\right)_{t\geq 0}$ and $(X_t^c)_{t\geq 0}$, where we recall that a right-continuous process $(Y_t)_{t\geq 0}$ is said to admit a fixed discontinuity at point $t > 0$ as soon as

$$\mathbb{P}(\{\omega \in \Omega : \Delta Y_t(\omega) \neq 0\}) > 0,$$

where $\Delta Y_t = Y_t - Y_{t^-}$.

As noted by [ÇIN 80], the process $\left(X_t^d\right)_{t\geq 0}$ has fixed discontinuities at exactly the same locations as the function A. Indeed, for $0 \leq s < t$, the random variable $X_t - X_s$ is gamma distributed $\mathcal{G}\left(A^d(t) - A^d(s), b\right)$ with

$$A^d(t) - A^d(s) = \sum_{u\in\mathcal{I}, s<u\leq t} \Delta A(u).$$

Based on this, the random variable $X_t^d - X_s^d$ is null (gamma distributed $\mathcal{G}(0, b)$) as soon as $\mathcal{I} \cap (s, t] = \varnothing$ and it is gamma distributed $\mathcal{G}(\Delta A(u), b)$ when $\mathcal{I} \cap (s, t] = \{u\}$. The process $\left(X_t^d\right)_{t\geq 0}$ is hence constant between the discontinuity points of A and for each discontinuity point t of A, the jump size ΔX_t^d is gamma distributed $\mathcal{G}(\Delta A(t), b)$. On the contrary, the function $A^c(t)$ is continuous, and the process $(X_t^c)_{t\geq 0}$ consequently has no fixed discontinuity. Based on theorem 12.1 of [DUD 11], the process $(X_t^c)_{t\geq 0}$ hence is stochastically continuous, namely:

$$\lim_{s\to t} \mathbb{P}(|X_t^c - X_s^c| > \varepsilon) = 0$$

for all $t \geq 0$ and all $\varepsilon > 0$.

EXAMPLE.– A few trajectories of $\left(X_t^d\right)_{0\leq t\leq 2}$, $(X_t^c)_{0\leq t\leq 2}$ and $(X_t)_{0\leq t\leq 2}$ are plotted in Figure 2.1 for

$$A(t) = t\, \mathbf{1}_{[0,0.5]} + \left(2 + t^2\right) \mathbf{1}_{(0.5,1]} + \left(4 + t^3\right) \mathbf{1}_{(1,1.5]} + \left(6 + t^4\right) \mathbf{1}_{(1.5,+\infty)}$$

and $b = 1$.

The stochastically discontinuous part $\left(X_t^d\right)_{t\geq 0}$ of $(X_t)_{t\geq 0}$ is easy to deal with and it does not bring much to the present study. We consequently go on only with the stochastically continuous part $(X_t^c)_{t\geq 0}$ and we drop exponent c. This means that from now on, the function A is assumed to be *continuous* (so that $(X_t)_{t\geq 0}$ is stochastically continuous).

2.4. Series representations

In all the following, we set $A : \mathbb{R}_+ \to \mathbb{R}_+$ to be a continuous and non-decreasing function, such that $A(0) = 0$. We also set $a(\mathrm{d}t)$ to be the only measure on $(\mathbb{R}_+, \mathcal{B}(\mathbb{R}_+))$ such that

$$A(t) = a([0, t]), \text{ for all } t \geq 0.$$

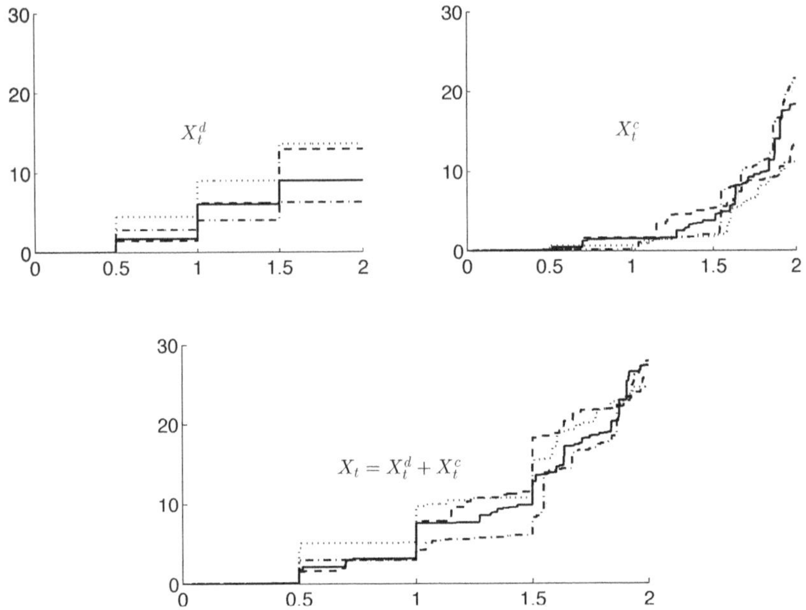

Figure 2.1. *A few trajectories of X_t^d, X_t^c and X_t on $[0, 2]$*

We recall that $a\,(\mathrm{d}t)$ is σ-finite and diffuse (see proposition 2.18 and its proof).

This section presents series representations for a gamma process, which allows us to understand the jump structure of a gamma process and also to provide tools for its simulation. These simulation tools are developed later on in section 2.5.3. Based on the scaling result from proposition 2.25, we only consider the case $b = 1$. Most of the results in this section are based on the study of Rosiński [ROS 01] (which is restricted to the homogeneous case however) (see also [BON 82] and [ÇIN 11]).

We begin with a technical classical result.

LEMMA 2.27.– Frullani integral. We have:

$$\int_{\mathbb{R}_+} \left(1 - e^{-sv}\right) \frac{e^{-v}}{v} \mathrm{d}v = \log\left(1 + s\right)$$

for all $s \geq 0$.

PROOF.– Let $f(s) = \int_{\mathbb{R}_+} (1 - e^{-sv}) \frac{e^{-v}}{v} \mathrm{d}v$, for all $s \geq 0$. As

$$\left| \frac{\partial}{\partial s} \left(\left(1 - e^{-sv}\right) \frac{e^{-v}}{v} \right) \right| = e^{-(s+1)v} \leq e^{-v}$$

with $\int_{\mathbb{R}_+} e^{-v} \mathrm{d}v < +\infty$, we can use differentiation under the integral sign to get

$$f'(s) = \int_{\mathbb{R}_+} \frac{\partial}{\partial s} \left(\left(1 - e^{-sv}\right) \frac{e^{-v}}{v} \right) \mathrm{d}v = \int_{\mathbb{R}_+} e^{-(s+1)v} \, \mathrm{d}v = \frac{1}{1+s}$$

which provides the result due to $f(0) = 0$. □

A direct consequence is that, for each $\alpha \geq 0$, we have

$$\exp\left(-\alpha \int_{\mathbb{R}_+} \left(1 - e^{-sv}\right) \frac{e^{-v}}{v} \mathrm{d}v\right) = \exp\left(-\alpha \,\log(1+s)\right) = \frac{1}{(1+s)^{\alpha}} \qquad [2.14]$$

for all $s \geq 0$, where we recognize the Laplace transform of the gamma distribution $\mathcal{G}(\alpha, 1)$ to be used later on.

We are now ready to provide a first series representation for a gamma process.

THEOREM 2.28.– Let $M = \sum_{n \geq 1} \delta_{(U_n, V_n)}$ be a Poisson (counting) random measure on $\mathbb{R}_+^2$ with intensity measure:

$$\nu(\mathrm{d}u, \mathrm{d}v) = a(\mathrm{d}u) \, \frac{e^{-v}}{v} \mathrm{d}v.$$

For each $t \geq 0$, we set $f_t(u, v) = v \, \mathbf{1}_{\{u \leq t\}}$ and

$$X_t = M f_t = \sum_{n \geq 1} f_t(U_n, V_n) = \sum_{n \geq 1} V_n \, \mathbf{1}_{\{U_n \leq t\}}.$$

Then, $(X_t)_{t \geq 0}$ is a gamma process $\mathcal{G}(A(t), 1)$.

PROOF.– We first check that $(X_t)_{t \geq 0}$ has independent increments. Let $0 \leq t_1 < \ldots < t_n$. Setting $g(u, v) = v$ and $B_i = (t_{i-1}, t_i] \times \mathbb{R}_+$ for each $2 \leq i \leq$

n, we have:

$$\begin{aligned} X_{t_i} - X_{t_{i-1}} &= M\left(f_{t_i} - f_{t_{i-1}}\right) \\ &= M\left(g\left(\mathbf{1}_{[0,t_i]\times\mathbb{R}_+} - \mathbf{1}_{[0,t_{i-1}]\times\mathbb{R}_+}\right)\right) \\ &= M\left(g\,\mathbf{1}_{B_i}\right) \\ &= M^{B_i}(g) \end{aligned}$$

using the notation of proposition 2.19. As the Borel sets B_i are pairwise disjoint ($2 \le i \le n$), the r.v.s $X_{t_2} - X_{t_1}$, ..., $X_{t_n} - X_{t_{n-1}}$ are independent so that $(X_t)_{t\ge 0}$ has independent increments. Furthermore, based on theorem 2.21 for Poïsson random measures, we have:

$$\begin{aligned} \mathbb{E}\left(e^{-s\left(X_{t_i}-X_{t_{i-1}}\right)}\right) &= \mathbb{E}\left(e^{-M^{B_i}(s\,g)}\right) \\ &= \exp\left(-\iint_{\mathbb{R}_+^2}\left(1-e^{-s\,g(u,v)}\right)\nu^{B_i}(\mathrm{d}u,\mathrm{d}v)\right) \\ &= \exp\left(-\int_{(t_{i-1},t_i]}\left(\int_{\mathbb{R}_+}\left(1-e^{-sv}\right)\frac{e^{-v}}{v}\mathrm{d}v\right)a(\mathrm{d}u)\right) \\ &= \exp\left(-\left(A(t_i)-A(t_{i-1})\right)\left(\int_{\mathbb{R}_+}\left(1-e^{-sv}\right)\frac{e^{-v}}{v}\mathrm{d}v\right)\right) \\ &= \left(\frac{1}{1+s}\right)^{A(t_i)-A(t_{i-1})} \end{aligned}$$

based on [2.14], where we recognize the Laplace transform of $\mathcal{G}\left(A(t_i) - A(t_{i-1}), 1\right)$ in the last line. As $(X_t)_{t\ge 0}$ is a right-continuous process with left-side limits, we conclude that it is a gamma process $\mathcal{G}(A(t), 1)$. □

The previous result shows that a gamma process is a pure jump process and that its jumps on $[0, t]$ are countably many:

$$X_t = \sum_{n\ge 1} V_n \mathbf{1}_{\{U_n \le t\}} = \sum_{s\le t} \Delta X_s \qquad [2.15]$$

where $\Delta X_s = X_s - X_{s^-}$, for all $s > 0$. For a jump at time U_n, the size of the jump is V_n.

Also, for each Borel set B, let $M_t(B)$ be the number of jumps on $(0,t]$ for which the jump size belongs to B :

$$\begin{aligned} M_t(B) &= \#\{s \in (0,t] : \Delta X_s \in B\} \\ &= \#\{n \geq 1 : U_n \leq t \text{ and } \Delta X_{U_n} = V_n \in B\} \end{aligned}$$

Then

$$\begin{aligned} M_t(B) &= \#\{n \geq 1 : (U_n, V_n) \in (0,t] \times B\} \\ &= M((0,t] \times B) \end{aligned}$$

and $M_t(B)$ is Poisson distributed with mean

$$\nu((0,t] \times B) = A(t) \times \int_B \frac{e^{-v}}{v} \mathrm{d}v$$

see definition 2.16.

Let $\varepsilon > 0$. Based on the fact that $\int_\varepsilon^{+\infty} \frac{e^{-v}}{v} \mathrm{d}v < +\infty$, the number of jumps of a gamma process $\mathcal{G}(A(t), b)$ on the time interval $[0,t]$ with jump size greater than ε is finite almost surely, with mean $A(t) \times \int_\varepsilon^{+\infty} \frac{e^{-v}}{v} \mathrm{d}v < +\infty$. Setting $M^{(\varepsilon)}$ to be a Poisson random measure on $\mathbb{R}_+^2$ with intensity measure:

$$\nu^{(\varepsilon)}(\mathrm{d}u, \mathrm{d}v) = a(\mathrm{d}u) \; \frac{e^{-v}}{v} \mathbf{1}_{[\varepsilon,\infty)} \mathrm{d}v,$$

the points (U_n, V_n) of $M^{(\varepsilon)}$ correspond to the jump times (U_n) and sizes (V_n) of a gamma process, restricted to the jumps of size greater than ε. It is then easy to generate an observation of these jump times and sizes on a bounded time interval using algorithm 2.23.

On the contrary, however small ε is, we have $\int_0^\varepsilon \frac{e^{-v}}{v} \mathrm{d}v = +\infty$ and the number of jumps of a gamma process on $[0,t]$ with size smaller than ε is almost surely infinite (but countable).

EXAMPLE.– We plot an observation of the points (U_n, V_n) of the Poisson random measure $M^{(\varepsilon)}$ with intensity measure $\mathrm{d}u \; \frac{e^{-v}}{v} \mathrm{d}v$ on $[0,T] \times [\varepsilon, +\infty)$ in Figure 2.2 with $\varepsilon = 10^{-10}$ using algorithm 2.23. The points (U_n, V_n) correspond to the jumps of a gamma process $(X_t)_{t \geq 0}$ with $A(t) = t$ (and $b = 1$) on $[0,T]$ with size greater than ε. Each U_n is a jump time and V_n is the size of the jump at time U_n. As expected, we observe that there is an accumulation of jumps of small size. However, note that large jumps are still possible.

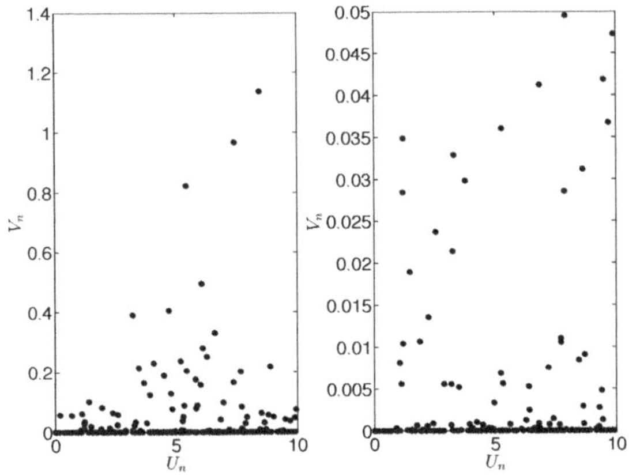

Figure 2.2. *An observation of the jump times (U_n) and jump sizes (V_n) of a gamma process $\mathcal{G}(t, 1)$, restricted to the jumps with size greater than 10^{-10} (left), and zoomed observation of the left plot, restricted to the jumps with size in $[10^{-10}, 0.05]$ (right)*

Based on the previous observations, Gamma processes are well adapted to model deterioration that is accumulating over time, as the result of many tiny increments [VAN 09].

REMARK 2.29.– A homogeneous Gamma process as a Lévy process. In the homogeneous case where $A(t) = at$ for all $t \geq 0$ with $a > 0$, we set

$$\mu(\mathrm{d}v) = a\frac{e^{-v}}{v}\mathrm{d}v.$$

The Laplace transform of X_t now is

$$\mathcal{L}_{X_t}(s) = \exp\left(-t\int_{\mathbb{R}_+}\left(1 - e^{-sv}\right)\mu(\mathrm{d}v)\right)$$

(see the proof of theorem 2.28). Based on Lévy-Kintchine formula for non-decreasing Lévy processes with null drift [BER 96], we derive that $\mu(\mathrm{d}u)$ is the *Lévy measure* of the homogeneous Gamma process $\mathcal{G}(at, b)$. The Lévy measure $\mu(\mathrm{d}u)$ is such that

$$\int_{\mathbb{R}_+} \mu(\mathrm{d}x)\,\mathrm{d}x = +\infty,$$

which is the characteristic of a Lévy process with *infinite activity*, namely such that "almost all paths have infinitely many jumps along any time interval of finite length" [EBE 09]. Also, we have $\int_{\mathbb{R}_+} (1 \wedge x)\,\mu(\mathrm{d}x) < +\infty$, which is characteristic of a Lévy process for which almost all paths have *bounded variation* [EBE 09].

The first series representation provided in theorem 2.28 allows to understand the jump structure of a Gamma process (in connection with its Lévy measure in the homogeneous case). However, this first series representation does not directly provide a way to simulate the paths of a Gamma process. Indeed, based on algorithm 2.23, it is easy to simulate the points $(U_n, V_n) \in [0,t] \times \mathbb{R}_+$ of a Poisson random measure with intensity measure of the shape $\nu(\mathrm{d}u, \mathrm{d}v) = a(\mathrm{d}u)\ \mu(\mathrm{d}v)$ when $\mu(\mathrm{d}v)$ is a bounded measure. Here, $\mu(\mathrm{d}v) = \frac{e^{-v}}{v}\mathrm{d}v$ is not bounded and we cannot directly construct the points $(U_n, V_n) \in [0,t] \times \mathbb{R}_+$. Because of this, we consider a slight modification of theorem 2.28 and we introduce

$$\mathrm{Ei}(x) = \int_x^{+\infty} \frac{e^{-v}}{v}\mathrm{d}v, \text{ for all } x > 0. \qquad [2.16]$$

We recall that Ei is the *exponential integral function*. The exponential integral function is a one-to-one decreasing function from $(0, +\infty)$ into $(0, +\infty)$. Setting $y = \mathrm{Ei}(x)$, it is easy to check that

$$\int_{\mathbb{R}_+} \left(1 - e^{-sx}\right) \frac{e^{-x}}{x}\mathrm{d}x = \int_{\mathbb{R}_+} \left(1 - e^{s\ \mathrm{Ei}^{-1}(y)}\right) \mathrm{d}y, \qquad [2.17]$$

which is the key of the following corollary.

COROLLARY 2.30.– [Inverse Lévy series representation]. Let $M = \sum_{n \geq 1} \delta_{(U_n, V_n)}$ be a Poisson random measure on $\mathbb{R}_+^2$ with intensity measure:

$$\nu(\mathrm{d}u, \mathrm{d}v) = \mathrm{d}u\, a(\mathrm{d}v).$$

For each $t \geq 0$, we set $f_t(u, v) = \mathrm{Ei}^{-1}(u)\ \mathbf{1}_{\{v \leq t\}}$ and

$$X_t = Mf_t = \sum_{n \geq 1} f_t(U_n, V_n) = \sum_{n \geq 1} \mathrm{Ei}^{-1}(U_n)\ \mathbf{1}_{\{V_n \leq t\}}.$$

Then, $(X_t)_{t \geq 0}$ is a gamma process $\mathcal{G}(A(t), 1)$.

PROOF.– Using similar arguments as for the proof of theorem 2.28, the only point to prove is that an increment $X_{t+h} - X_t$ (with $h, t \geq 0$) is gamma distributed $\mathcal{G}(A(t+h) - A(t), 1)$. For each $s > 0$, we have:

$$\begin{aligned}
\mathbb{E}\left(e^{-s(X_{t+h}-X_t)}\right) &= e^{-M(s(f_{t+h}-f_t))} \\
&= \exp\left(-\iint_{\mathbb{R}_+^2}\left(1 - e^{-s(f_{t+h}-f_t)}\right) \mathrm{d}\nu\right) \\
&= \exp\left(-\int_{(t,t+h]}\left(\int_{\mathbb{R}_+}\left(1 - e^{s\ \mathrm{Ei}^{-1}(y)}\right) \mathrm{d}y\right) a(\mathrm{d}v)\right) \\
&= \exp\left((A(t+h) - A(t)) \int_{\mathbb{R}_+}\left(1 - e^{-sx}\right) \frac{e^{-x}}{x} \mathrm{d}x\right)
\end{aligned}$$

based on [2.17]. This allows us to conclude in the same way as for theorem 2.28.

□

From a numerical point of view, this inverse Lévy representation is not very well adapted for the simulation of gamma processes, because it requires the computation of the inverse exponential integral function (see section 2.5.3 for more details). We consequently go on with three other similar results, which are better adapted for simulation purpose (see [ROS 01] in the homogeneous case).

THEOREM 2.31.– Let $m(\mathrm{d}w)$ be a probability measure on $(\mathbb{R}_+, \mathcal{B}(\mathbb{R}_+))$ and let $M = \sum_{n \geq 1} \delta_{(U_n, V_n, W_n)}$ be a Poisson (counting) random measure on $\mathbb{R}_+^3$ with intensity measure:

$$\nu(\mathrm{d}u, \mathrm{d}v, \mathrm{d}w) = \mathrm{d}u\ a(\mathrm{d}v)\ m(\mathrm{d}w).$$

Let $g : \mathbb{R}_+^2 \to \mathbb{R}_+$ be a measurable function and, for $t \geq 0$, let $f_t(u,v,w) = g(u,w)\, \mathbf{1}_{[0,t]}(v)$ and

$$X_t = Mf_t = \sum_{n \geq 1} g(U_n, W_n)\, \mathbf{1}_{[0,t]}(V_n).$$

Then, for each of the following choices for $m(\mathrm{d}w)$ and $g(u,w)$, the process $(X_t)_{t \geq 0}$ is a gamma process $\mathcal{G}(A(t), 1)$.

1) *Bondesson's series representation* [BON 82]: $g(u,w) = w\, e^{-u}$ and $m(\mathrm{d}w) = e^{-w}\mathrm{d}w$ (standard exponential distribution).

2) *Rejection series representation*: $g(u,w) = \frac{1}{\exp(u)-1}\mathbf{1}_E(u,w)$ with

$$E = \left\{ (u,w) \in \mathbb{R}_+^* \times [0,1] : \frac{e^u}{e^u - 1} \exp\left(-(e^u - 1)^{-1}\right) \geq w \right\}$$

and $m(\mathrm{d}w) = \mathbf{1}_{[0,1]}(w)\,\mathrm{d}w$ (standard uniform distribution).

3) *Thinning series representation*: $g(u,w) = w\, \mathbf{1}_{\{uw \leq 1\}}$ and $m(\mathrm{d}w) = e^{-w}\mathrm{d}w$ (standard exponential distribution).

PROOF.– Using similar arguments as for the proof of theorem 2.28, it is enough to compute the Laplace transform of $X_{t+h} - X_t$, with $t, h \geq 0$ for each of the three cases. For $s > 0$, we have:

$$\begin{aligned}
\mathbb{E}\left(e^{-s(X_{t+h}-X_t)}\right) &= \exp\left(-\iiint_{\mathbb{R}_+^3} \left(1 - e^{-s(f_{t+h}-f_t)}\right) d\nu\right) \\
&= \exp\left(-\int_{(t,t+h]} \left(\iint_{\mathbb{R}_+^2} \left(1 - e^{-s\, g(u,w)}\right) \mathrm{d}u\, m(\mathrm{d}w)\right) a(\mathrm{d}v)\right) \\
&= \exp\left(-\left(A(t+h) - A(t)\right) G(s)\right) \qquad [2.18]
\end{aligned}$$

where

$$G(s) = \iint_{\mathbb{R}_+^2} \left(1 - e^{-s\, g(u,w)}\right) \mathrm{d}u\, m(\mathrm{d}w).$$

Following the proof of theorem 2.28, the only point is to check that

$$G(s) = \int_{\mathbb{R}_+} \left(1 - e^{-sx}\right) \frac{e^{-x}}{x} dx$$

for each of the following three cases.

1) *Bondesson*: we have

$$G(s) = \iint\limits_{\mathbb{R}_+^2} \left(1 - e^{-s\, w\, e^{-u}}\right) e^{-w} du\, dw.$$

Setting $(x, y) = (w\, e^{-u}, e^{u})$, we get

$$G(s) = \iint\limits_{\mathbb{R}_+ \times [1, +\infty)} \left(1 - e^{-s\, x}\right) e^{-xy} dx\, dy = \int_{\mathbb{R}_+} \left(1 - e^{-s\, x}\right) \frac{e^{-x}}{x}\, dx.$$

2) *Rejection*: we have

$$G(s) = \iint\limits_{\mathbb{R}_+^* \times [0,1]} \left(1 - e^{-s\, \frac{1}{\exp(u) - 1} \mathbf{1}_E(u, w)}\right) du\, dw.$$

Setting $(x, y) = \left(\frac{1}{\exp(u) - 1}, w\right)$ and noting that $(u, w) \in E$ if and only if $y \leq (1 + x)\, e^{-x}$ (≤ 1), we get:

$$G(s) = \int_0^{+\infty} \left(\int_0^{(1+x)e^{-x}} dy\right) \left(1 - e^{-s\, x}\right) \frac{1}{x\,(1 + x)}\, dx$$
$$= \int_0^{+\infty} \left(1 - e^{-s\, x}\right) \frac{e^{-x}}{x}\, dx.$$

3) *Thinning*: we have

$$G(s) = \iint\limits_{\mathbb{R}_+^2} \left(1 - e^{-sw\mathbf{1}_{\{uw \leq 1\}}}\right) e^{-w} du\, dw$$
$$= \int_{\mathbb{R}+} \left(1 - e^{-sw}\right) \left(\int_0^{\frac{1}{w}} du\right) e^{-w} dw$$
$$= \int_{\mathbb{R}+} \left(1 - e^{-sw}\right) \frac{e^{-w}}{w} dw.$$

□

2.5. Simulation

The point here is to see how to simulate the paths of a gamma process $(Y_t)_{t\geq 0}$ with distribution $\mathcal{G}(A(t), b)$, where $b > 0$. Based on proposition 2.25, it is enough to see how to simulate the paths of a gamma process $(X_t)_{t\geq 0}$ with distribution $\mathcal{G}(A(t), 1)$ and next set $Y_t = \frac{X_t}{b}$ for all $t \geq 0$. Also, in this section, T is a fixed horizon time ($T > 0$) and the specific point is to simulate the paths of $(X_t)_{t\geq 0}$ on the bounded set $[0, T]$. As $(X_t)_{t\in[0,T]}$ almost surely jumps infinitely many on $[0, T]$ (see the lines following theorem 2.28), it is impossible to simulate exact trajectories and the very point is to propose methods to simulate approximate paths of $(X_t)_{t\in[0,T]}$ or exact paths observed at given fixed points.

2.5.1. *Random walk approximation*

Let $(t_i)_{0\leq i\leq N}$ with $t_0 = 0 < t_1 < \cdots < t_i < t_{i+1} < \cdots < t_N = T$. The random walk approximation (RWA) method allows us to exactly simulate $(X_{t_i})_{0\leq i\leq N}$ at the fixed grid points $(t_i)_{0\leq i\leq N}$. For each $n \in \{1, \cdots, N\}$, we simply observe that

$$X_{t_n} = \sum_{i=0}^{n-1} \left(X_{t_{i+1}} - X_{t_i}\right).$$

The increments $X_{t_{i+1}} - X_{t_i}$ are independently generated according to $\mathcal{G}(A(t_{i+1}) - A(t_i), 1)$ for all $0 \leq i \leq N - 1$ and next summed up to provide X_{t_n}, for all $1 \leq n \leq N$. Based on corollary 2.3, this means that X_{t_n} is exactly generated according to the gamma distribution $\mathcal{G}\left(\sum_{i=0}^{n-1} (A(t_{i+1}) - A(t_i)), 1\right) = \mathcal{G}(A(t_n), 1)$.

ALGORITHM 2.32.– Generation of an observation $(x_0, \cdots, x_N)$ of $(X_{t_i})_{0\leq i\leq N}$ by Random Walk Approximation:

– for $i \in \{0, \cdots, N-1\}$, simulate independent observations Δx_i according to $\mathcal{G}(A(t_{i+1}) - A(t_i), 1)$;

– set $x_0 = 0$ and for $n \in \{1, \cdots, N\}$, set $x_n = \sum_{i=0}^{n-1} \Delta x_i$.

Based on the exactly simulated $(X_{t_i})_{0\leq i\leq N}$ at the grid points, an approximate trajectory for $(X_t)_{t\in[0,T]}$ is next obtained through the following piecewise constant path:

$$X_t^{(N)} = \sum_{i=0}^{N-1} X_{t_i}\, \mathbf{1}_{[t_i,t_{i+1})}(t) \text{ for all } t \in [0,T). \qquad [2.19]$$

Note that this provides a lower approximation, with $X_t^{(N)} \leq X_t$. Other possibilities would be to consider an upper approximation $\left(X_t^{(N,+)} \geq X_t\right)$ with

$$X_t^{(N,+)} = \sum_{i=0}^{N-1} X_{t_{i+1}}\, \mathbf{1}_{(t_i,t_{i+1}]}(t) \text{ for all } t \in (0,T]$$

or some "middle point" approximation

$$X_t^{(N,\text{middle point})} = \sum_{i=0}^{N-1} \frac{X_{t_i}+X_{t_{i+1}}}{2}\, \mathbf{1}_{[t_i,t_{i+1})}(t) \text{ for all } t \in [0,T)$$

or some "piecewise linear" approximation

$$X_t^{(N,\text{piecewise linear})} = \sum_{i=0}^{N-1} \left(X_{t_i} + \frac{X_{t_{i+1}}-X_{t_i}}{t_{i+1}-t_i}(t-t_i)\right) \mathbf{1}_{[t_i,t_{i+1})}(t)$$
$$\text{for all } t \in [0,T).$$

Formula [2.19] is the most commonly used and we consequently go on with this definition of the approximate process. We now study the error between X_t and $X_t^{(N)}$ in case of a regular grid.

PROPOSITION 2.33.– For $n \in \mathbb{N}^*$, let $X_t^{(N)}$ be defined by [2.19] where $t_i = i\,\frac{T}{N}$. Then, $X_t - X_t^{(N)}$ is non-negative and gamma distributed $\mathcal{G}\left(A(t) - A\left(\frac{T}{N}\left\lfloor t\,\frac{N}{T}\right\rfloor\right)\right)$. In case $A(t)$ is Lipschitz continuous on $[0,T]$

with Lipschitz constant K_T, then:

$$\mathbb{E}\left(X_t - X_t^{(N)}\right) \le K_T \frac{T}{N},$$

$$\mathrm{Var}\left(X_t - X_t^{(N)}\right) \le K_T \frac{T}{N}$$

for all $t \in [0, T]$.

PROOF.– For $t_i \le t < t_{i+1}$, the random variable

$$X_t - X_t^{(N)} = X_t - X_{t_i}$$

is gamma distributed $\mathcal{G}\left(A\left(t\right) - A\left(t_i\right)\right)$. This provides the first result, noting that $t_i \le t < t_{i+1}$ is equivalent to $i\, \frac{T}{N} \le t < \left(i+1\right) \frac{T}{N}$ namely to $i = \left\lfloor t\, \frac{N}{T} \right\rfloor$. If $A\left(t\right)$ is Lipschitz continuous on $[0, T]$, we get:

$$\begin{aligned}\mathbb{E}\left(X_t - X_t^{(N)}\right) &= \mathrm{Var}\left(X_t - X_t^{(N)}\right)\\ &= A\left(t\right) - A\left(\frac{T}{N}\left\lfloor t\,\frac{N}{T}\right\rfloor\right)\\ &\le K_T\left(t - \frac{T}{N}\left\lfloor t\,\frac{N}{T}\right\rfloor\right)\\ &\le K_T \frac{T}{N}.\end{aligned}$$

□

As for all the other forthcoming simulation procedures, numerical experiments will be provided in section 2.5.4. This will allow us to compare one method to the other.

As noted by Rosinski [ROS 08], "a disadvantage of such method [RWA] is that one cannot precisely identify the location and magnitude of the large jumps". Besides, when thinking of application to reliability, an important random variable is the time when the process first exceeds a given threshold (the failure threshold), which stands for the time to failure (see section 2.6 for more details). As the process evolves only through jumps, a time to failure corresponds to a jump time. Also, when the threshold is exceeded at the time of a large jump (see Figure 2.2), the threshold may be largely exceeded that may lead to a more severe failure. For our applicative purposes, it is hence important to catch "the location and the magnitude of the large jumps" through the approximate simulation procedure. As a result, even if the RWA is probably the simplest and the most commonly used method in reliability engineering for simulating gamma processes, and even if it provides exact simulation at given fixed points, it may not be always the best adapted method when dealing with hitting times.

2.5.2. *Gamma bridge sampling*

Just as for the previous RWA method, the gamma bridge sampling (GBS) method [DUF 91, AVR 03] allows us to exactly simulate $(X_{t_i})_{0\leq i\leq N}$ at given points $(t_i)_{0\leq i\leq N}$, which provides a piecewise constant approximate path for $(X_t)_{t\in[0,T]}$, accordingly.

The GBS method relies on the following remark: for each $0 \leq s < t < u$, let us write

$$X_t - X_s = \frac{X_t - X_s}{X_u - X_s}(X_u - X_s) = \frac{U}{U+V}(U+V)$$

where $U = X_t - X_s$ and $V = X_u - X_t$. Based on proposition 2.6, we can observe that $Y = U/(U+V) = (X_t - X_s)/(X_u - X_s)$ and $X_u - X_s = U+V$ are independent, where Y is beta distributed $\mathcal{B}(A(t) - A(s), A(u) - A(t))$. As a result, given that $X_s = x_s$ and $X_u = x_u$, then $X_t = X_s + (X_t - X_s)$ is conditionally identically distributed as

$$x_s + Y(x_u - x_s),$$

where Y is beta distributed $\mathcal{B}(A(t) - A(s), A(u) - A(t))$. This fact allows us to simulate an observation of X_t, given an observation of (X_s, X_u).

Taking $N = 2^n$ and $t_i = \frac{i}{2^n}T$ for all $0 \leq i \leq 2^n$, and starting from $X_0 = 0$ and from an observation of X_T, the previous remark allows us to construct an observation of $X_{T/2}$. Next, starting from $(X_0, X_{T/2})$ and from $(X_{T/2}, X_T)$, we can construct observations of $X_{T/4}$ and $X_{3T/4}$, respectively, and so on.

More generally, taking $N = 2^n$ and $t_0 = 0 < t_1 < \cdots < t_i < t_{i+1} < \cdots < t_{2^n} = T$, the method allows us to recursively construct an observation of $(X_{t_i})_{0\leq i\leq 2^n}$ with the following steps, setting $Z_i = X_{t_i}$ for sake of simplicity. Step 0: construction of (Z_0, Z_{2^n}). Step 1: construction of $Z_{2^{n-1}}$ from (Z_0, Z_{2^n}). Step 2: construction of $Z_{(2\times 0+1)2^{n-2}}$ from $(Z_0, Z_{2^{n-1}})$ and construction of $Z_{(2\times 1+1)2^{n-2}}$ from $(Z_{2^{n-1}}, Z_{2^n})$. Step i: construction of $Z_{(2j+1)\,2^{n-i}}$ from $(Z_{j\,2^{n-i+1}}, Z_{(j+1)\,2^{n-i+1}})$ for each $0 \leq j \leq 2^{i-1} - 1$. Step n: construction of Z_{2j+1} from $\left(Z_{j\,2}, Z_{(j+1)\,2}\right)$ for each $0 \leq j \leq 2^{n-1} - 1$. This provides the following algorithm.

ALGORITHM 2.34.– Generation of an observation $(x_0, \cdots, x_{2^n})$ of $(X_{t_i})_{0 \leq i \leq 2^n}$ by Gamma Bridge Sampling:

– initialization: set $x_0 = 0$ and generate an observation x_{2^n} according to $\mathcal{G}\left(A\left(t_{2^n}\right), 1\right)$;

– for $i = 1, \cdots, n$ do

- for $j = 0, \cdots, 2^{i-1} - 1$ do

- generate an observation y_j according to

$$\mathcal{B}\left(A\left(t_{(2j+1)\,2^{n-i}}\right) - A\left(t_{j\,2^{n-i+1}}\right), A\left(t_{(j+1)\,2^{n-i+1}}\right) - A\left(t_{(2j+1)\,2^{n-i}}\right)\right);$$

- set $x_{(2j+1)\,2^{n-i}} = x_{j\,2^{n-i+1}} + y_j\ \left(x_{(j+1)\,2^{n-i+1}} - x_{j\,2^{n-i+1}}\right)$,

- next j (EndFor),

– next i (EndFor),

All the comments made on the RWA method merely remain valid for the GBS method. However, whereas in the RWA method, X_T is sampled as the sum of eventually many tiny increments leading to some possible round-off errors; in the GBS method, X_T is the first to be sampled, next $X_T/2$, and so on. We may consequently expect the GBS method to be more accurate for these points.

2.5.3. *Compound Poisson process approximations*

Compound Poisson process approximations are obtained by truncating the series representations from section 2.4. As already noted, the inverse Lévy representation provided by corollary 2.30 requires the computation of the inverse exponential integral function Ei^{-1}, which is time consuming. We consequently prefer to restrict ourselves to the three other series representations provided by theorem 2.31. The interested reader may however consult [IMA 13] for an in-depth account on the use of the inverse Lévy representation for simulation purposes.

Recall that each of the three representations of theorem 2.31 may be written as

$$X_t = Mf_t = \sum_{n\geq 1} g\left(U_n, W_n\right) \mathbf{1}_{[0,t]}\left(V_n\right) \text{ for all } t \geq 0 \qquad [2.20]$$

where $f_t\left(u,v,w\right) = \sum_{n\geq 1} g\left(u,w\right) \mathbf{1}_{[0,t]}\left(v\right)$, $M = \sum_{n\geq 1} \delta_{\left(U_n,V_n,W_n\right)}$ is a Poisson random measure on $\mathbb{R}_+^3$ with intensity measure:

$$\nu\left(\mathrm{d}u, \mathrm{d}v, \mathrm{d}w\right) = \mathrm{d}u\, a\left(\mathrm{d}v\right)\ m\left(\mathrm{d}w\right)$$

and where $m\left(\mathrm{d}w\right)$ is a probability measure on $\left(\mathbb{R}_+, \mathcal{B}\left(\mathbb{R}_+\right)\right)$ (with different g and $m\left(\mathrm{d}w\right)$ for the three series representations).

As the point is to sample $\left(X_t\right)_{0\leq t\leq T}$ on $[0,T]$ with T fixed, it is enough to construct only the points $\left(U_n, V_n, W_n\right) \in \mathbb{R}_+ \times [0,T] \times \mathbb{R}_+$. These points however are infinitely many and it is necessary to truncate the series representation [2.20] to get a usable expression in practice. With this aim, we retain only the points $\left(U_n, V_n, W_n\right) \in [0,K] \times [0,T] \times \mathbb{R}_+$, with $K > 0$. This truncation amounts to keeping only the terms $g\left(U_n, W_n\right)$ of the series such that $U_n \leq K$ (and $V_n \leq T$). Based on the fact that, for the three series representations, $g\left(u,w\right)$ is decreasing with respect of u on its support, this means that the truncation boils down to remove the jumps of small size, with size below $g\left(K, W_n\right)$. Once the points $\left(U_n, V_n, W_n\right) \in [0,K] \times [0,T] \times \mathbb{R}_+$ are constructed, the approximate path $\left(X_t^{(K)}\right)_{0\leq t\leq T}$ is next constructed through

$$X_t^{(K)} = M\left(f_t\, \mathbf{1}_{[0,K]\times[0,T]\times\mathbb{R}_+}\right) = \sum_{n\geq 1} g\left(U_n, W_n\right) \mathbf{1}_{[0,K]}\left(U_n\right) \mathbf{1}_{[0,t]}\left(V_n\right)$$

for all $0 \leq t \leq T$.

The point now is to see how to construct the points $\left(U_n, V_n, W_n\right) \in [0,K] \times [0,T] \times \mathbb{R}_+$. This amounts to constructing the points of a Poisson random measure $M_T^{(K)}$ with intensity measure:

$$\begin{aligned}\nu_T^{(K)}\left(\mathrm{d}u, \mathrm{d}v, \mathrm{d}w\right) &= \mathbf{1}_{[0,K]}\left(u\right) \mathrm{d}u\, \mathbf{1}_{[0,T]}\left(v\right) a\left(\mathrm{d}v\right)\ m\left(\mathrm{d}w\right)\\ &= \left(A\left(T\right) \mathbf{1}_{[0,K]}\left(u\right)\ \mathrm{d}u\right) \left(\mathbf{1}_{[0,T]}\left(v\right) \frac{a\left(\mathrm{d}v\right)}{A\left(T\right)}\right)\ m\left(\mathrm{d}w\right).\end{aligned}$$

Based on corollary 2.22, the generation of the corresponding points is easy, generating first the U_n's as the points on $[0, K]$ of a homogeneous Poisson process with rate $A(T)$, and next independently generating the V_n's and W_n's as independent r.v.s with distribution $\mathbf{1}_{[0,T]}(v)\,\frac{a(dv)}{A(T)}$ and $m(dw)$, respectively. Mimicking algorithm 2.23, we get the following algorithm (assuming $A(t)$ to be one-to-one).

ALGORITHM 2.35.– Generation of an approximate path of $(X_t)_{t\geq 0}$ through truncated series representation:

1) Generate an observation n according to the Poisson distribution $\mathcal{P}(A(T)\,K)$.

2) Generate n i.i.d. observations $x_1, \cdots, x_n$ according to the standard uniform distribution $\mathcal{U}([0,1])$ (independently of n) and set $u_i = Kx_i$ for $i = 1, \cdots, n$ (these are observations of $U_i \in [0, K]$ with $1 \leq i \leq n$).

3) Generate n i.i.d. observations $y_1, \cdots, y_n$ according to the standard uniform distribution (independently of n and of the u_i's) and set $v_i = A^{-1}(A(T)\,y_i)$ for $i = 1, \cdots, n$ (these are observations of $V_i \in [0, T]$ with $1 \leq i \leq n$).

4) Generate n i.i.d. observations $w_1, \cdots, w_n$ with distribution $m(dw)$ (independently of n, u_i's and v_i's), where $m(dw)$ is:

- the standard exponential distribution (with mean 1) for Bondesson's and the thinning series representations;

- the standard uniform distribution $\mathcal{U}([0,1])$ for the rejection series representation.

These are observations of W_i with $1 \leq i \leq n$.

5) Set $x_t = \sum_{n\geq 1} g(u_n, w_n)\,\mathbf{1}_{[0,t]}(v_n)$ for each $t \in [0, T]$, where:

- $g(u, w) = w\, e^{-u}$ for the Bondesson's series representation;

- $g(u, w) = \frac{1}{\exp(u)-1}\mathbf{1}_{\left\{\frac{e^u}{e^u-1}\exp(-(e^u-1)^{-1})\geq w\right\}}$ for the rejection series representation;

- $g(u, w) = w\,\mathbf{1}_{\{uw\leq 1\}}$ for the thinning series representation.

Note that the sorting step of algorithm 2.10 of the points $u_1, \cdots, u_n$ of the Poisson process is here useless. Also, in the homogeneous case $A(t) = at$, the second part of step 3 simply becomes $v_i = T\,y_i$.

Writing

$$\nu_T^{(K)}(\mathrm{d}u, \mathrm{d}v, \mathrm{d}w) = \left(\frac{1}{K}\mathbf{1}_{[0,K]}(u)\ \mathrm{d}u\right)\left(K\ \mathbf{1}_{[0,T]}(v)\ a(\mathrm{d}v)\right)\ m(\mathrm{d}w),$$

another (equivalent) algorithm is also possible, constructing first the V_n's as the points on $[0,T]$ of a non-homogeneous Poisson process $\left(N_t^{(K)}\right)_{t\geq 0}$ with intensity measure $K\ a(\mathrm{d}v)$, and next the $U_n^{(K)}$'s (independently) according to the uniform distribution on $[0,K]$ and finally the $W_n^{(K)}$'s (independently) with distribution $m(\mathrm{d}u)$. This construction provides

$$X_t^{(K)} = \sum_{n=1}^{N_t^{(K)}} g\left(U_n^{(K)}, W_n^{(K)}\right)$$

and allows us to interpret the truncated series as a compound Poisson process.

We now study the impact of the truncation in the series representation [2.20].

PROPOSITION 2.36.– Let

$$R_t^{(K)} = X_t - X_t^{(K)}$$

be the sum of the omitted terms in the series representation [2.20]. Then, $\left(R_t^{(K)}\right)_{t\geq 0}$ is a non-negative process with independent increments. Also:

1) *Bondesson's series representation*: $\left(R_t^{(K)}\right)_{t\geq 0}$ is a gamma process $\mathcal{G}\left(A(t), e^K\right)$ with $\mathbb{E}\left(R_t^{(K)}\right) = A(t)\,e^{-K}$ and $\mathrm{Var}\left(R_t^{(K)}\right) = A(t)\,e^{-2K}$.

2) *Rejection series representation*:

$$\mathbb{E}\left(R_t^{(K)}\right) = A(t)\left(1 - \exp\left(-\frac{1}{e^{K-1}}\right)\right) \underset{K\to+\infty}{\sim} A(t)\,e^{-K}$$

and

$$\mathrm{Var}\left(R_t^{(K)}\right) = A(t)\left(\frac{1}{e^{K-1}}\exp\left(-\frac{1}{e^{K-1}}\right) - 1 + \exp\left(-\frac{1}{e^{K-1}}\right)\right) \underset{K\to+\infty}{\sim} A(t)\,\frac{e^{-2K}}{2}$$

where $\underset{K\to+\infty}{\sim}$ means "equivalent to" when $K \to +\infty$.

3) *Thinning series representation*:

$$\mathbb{E}\left(R_t^{(K)}\right) = A(t)\left(1 - K\left(1 - e^{-\frac{1}{K}}\right)\right) \underset{K\to+\infty}{\sim} A(t)\frac{1}{2K}$$

and

$$\mathrm{Var}\left(R_t^{(K)}\right) = A(t)\left(1 + e^{-\frac{1}{K}} - 2K\left(1 - e^{-\frac{1}{K}}\right)\right) \underset{K\to+\infty}{\sim} A(t)\frac{1}{6K^2}.$$

PROOF.– We have $R_t^{(K)} = X_t - X_t^{(K)} = M\left(f_t \mathbf{1}_{(K,+\infty)\times\mathbb{R}_+\times\mathbb{R}_+}\right)$. Noting that

$$\mathcal{L}_{R_t^{(K)}}(s) = \mathbb{E}\left(e^{-sR_t^{(K)}}\right) = \mathbb{E}\left(e^{-M\left(sf_t\mathbf{1}_{(K,+\infty)\times\mathbb{R}_+\times\mathbb{R}_+}\right)}\right)$$

for all $s > 0$, we calculate $\mathcal{L}_{R_t^{(K)}}(s)$ with the help of theorem 2.21, from where we derive the previous expressions for $\mathbb{E}\left(R_t^{(K)}\right)$ and $\mathrm{Var}\left(R_t^{(K)}\right)$, using the same changes in variables as for the proof of theorem 2.31. □

2.5.4. *Numerical experiments*

For both RWA and GBS, approximate trajectories are generated on the basis of the same regular grid (the *generation* grid) with step size $h = \frac{T}{2^{n_0}}$. We already know that both RWA and GBS provide exact simulation methods at the points of this generation grid. The results are tested at a few points of this generation grid, which are gathered into a grid denoted by $\mathcal{T}^{\text{gen}}$ in the following. The results are also tested at points of another grid $\mathcal{T}^{\text{test}}$ (a *test* grid) to control the quality of the simulation out of the regular grid used for their generation.

As for the truncated series representation, they do not require any grid to be fixed in advance. They are compared one to the other on a test grid $\mathcal{T}^{\text{test}}$. They are next compared to RWA and GBS both at points of the generation grid for RWA and GBS (on $\mathcal{T}_{n_0}^{\text{gen}}$) and on a test grid $\mathcal{T}^{\text{test}}$.

For each simulation method, we generate 1,000 independent sets of 1,000 independent approximate trajectories. Setting $\mathcal{T} = (t_i)_{1\leq i\leq n_1}$ to be a given grid (with points eventually taken among the generation grid of RWA and

GBS), for each $1 \leq i \leq n_1$, we have at our disposal 1,000 sets of 1,000 i.i.d. observations of X_{t_i}. For each $1 \leq i \leq n_1$ and for each set of 1,000 observations of X_{t_i}, we test the adequation of the observations to the gamma distribution $\mathcal{G}(A(t_i), 1)$ with the help of a Kolmogorov–Smirnov test (null assumption $\mathcal{H}_0$ = gamma distribution $\mathcal{G}(A(t_i), 1)$) at the significance level $\alpha = 0.05$. For each of these tests, we compute the p-value, where we recall that a value larger than 0.1 indicates that there is no presumption against the null assumption. We also compute the test statistic, which should be below a critical value for the null assumption not to be rejected. This critical value is called the *cut-off* value in the following. For each $1 \leq i \leq n_1$, we next compute the proportion of times the null assumption is rejected among the 1,000 tests (the coverage probability), which should be around 0.05, as well as the mean of 1,000 computed p-values and both the mean and standard deviation of 1,000 test statistics.

In all the following, we consider $T = 1$.

We first take $A(t) = t$ and $n_0 = 6$ (which controls the step size of the generation grid for RWA and GBS methods), with the following test grid taken from the generation points:

$$\begin{aligned}
&\mathcal{T}^{\text{gen}} \\
&= 2^{-6} \times [7, 13, 19, 25, 31, 37, 43, 49, 55, 61] \\
&\simeq [0.1094, 0.2031, 0.2969, 0.3906, 0.4844, 0.5781, 0.6719, \\
&\quad 0.7656, 0.8594, 0.9531].
\end{aligned}$$

The results are provided for RWA and GBS methods in Table 2.1. As expected, the results are pretty good for both methods at points of the grid used for their generation.

We next keep $A(t) = t$ and take

$$\mathcal{T}^{test} = [0.105, 0.200, 0.295, 0.390, 0.485, 0.580, 0.675, 0.770, 0.865, 0.960]$$

as test grid, where the points do not belong to the generation grid.

The results are provided for both RWA and GBS methods in Table 2.2 for $n_0 = 6$ and Table 2.3 for $n_0 = 9$. As can be seen, the results are not very good for small values of t_i and $n_0 = 6$. As expected, the results are improving when the generating grid is refined by increasing n_0. For instance, though the corresponding tables are not provided here, for $t_3 = 0.2950$, the proportion of time $\mathcal{H}_0$ is rejected is 0.0710 (RWA) and 0.0840 (GBS) for $n_0 = 7$, and 0.0440 (RWA) and 0.0640 (GBS) for $n_0 = 8$. However, for $t_1 = 0.1050$ and $n_0 = 8$, it is still 0.1320 (RWA) and 0.1420 (GBS). It is hence necessary to go up to $n_0 = 9$ to get correct results at all points of the test grid, which induces long computational times.

t_i	Coverage probability		p-value		Test statistic	
	RWA	GBS	RWA	GBS	RWA	GBS
0.1094	0.0510	0.0390	0.4980	0.5162	0.0274 (0.0082)	0.0268 (0.0077)
0.2031	0.0500	0.0390	0.4958	0.5064	0.0273 (0.0080)	0.0270 (0.0078)
0.2969	0.0480	0.0390	0.5065	0.4934	0.0272 (0.0084)	0.0273 (0.0080)
0.3906	0.0460	0.0370	0.5040	0.4945	0.0271 (0.0080)	0.0273 (0.0079)
0.4844	0.0460	0.0460	0.5093	0.4987	0.0270 (0.0080)	0.0272 (0.0080)
0.5781	0.0390	0.0440	0.5002	0.4975	0.0272 (0.0078)	0.0273 (0.0082)
0.6719	0.0460	0.0480	0.4914	0.4982	0.0275 (0.0081)	0.0273 (0.0082)
0.7656	0.0530	0.0520	0.4896	0.4949	0.0276 (0.0082)	0.0275 (0.0082)
0.8594	0.0510	0.0510	0.5064	0.5034	0.0272 (0.0082)	0.0272 (0.0083)
0.9531	0.0560	0.0520	0.4978	0.4946	0.0274 (0.0084)	0.0275 (0.0082)
Mean	0.0486	0.0447	0.4999	0.4998	0.0273 (0.0081)	0.0273 (0.0080)

Table 2.1. *Coverage probability, p-value and mean (std) of the test statistic for grid $\mathcal{T}^{gen}$ and $n_0 = 6$; cpu time = 228.9159; cutoff value = 0.0428*

Based on these tables (and on others not provided here), we can see that both RWA and GBS provide very similar results. As RWA is simpler with much lower computing times, we only retain RWA method for the sequel.

Going on with $A(t) = t$, the results are next provided on the same test grid $\mathcal{T}^{test}$ for the Bondesson, Rejection and Thinning methods in Tables 2.4 and 2.5 for $K = 40$ (cpu time for both tables: 249.71). Based on these results (and on others not provided here), it can be seen that both Bondesson's and the Rejection methods provide good results for a homogeneous gamma process as soon as $K \geq 40$. As for the Thinning method, it would require much higher K and consequently much higher computational times, as was expected from proposition 2.36. *We consequently do not consider the Thinning method any more in the following.* As for the comparison between Bondesson's and the Rejection methods, the computational times and the results are very similar for both. We consequently retain only one and *we choose to go on with Bondesson's method in the following (but we could have chosen Rejection method).*

t_i	Coverage probability		p-value		Test statistic	
	RWA	GBS	RWA	GBS	RWA	GBS
0.1050	0.8140	0.7940	0.0420	0.0446	0.0549 (0.0136)	0.0548 (0.0138)
0.2000	0.3660	0.3900	0.2067	0.2033	0.0392 (0.0120)	0.0396 (0.0122)
0.2950	0.2390	0.2360	0.2994	0.2974	0.0349 (0.0114)	0.0349 (0.0114)
0.3900	0.1920	0.1800	0.3522	0.3463	0.0328 (0.0109)	0.0330 (0.0111)
0.4850	0.0520	0.0500	0.4984	0.4800	0.0273 (0.0081)	0.0277 (0.0079)
0.5800	0.0460	0.0580	0.5017	0.4866	0.0273 (0.0083)	0.0277 (0.0083)
0.6750	0.0500	0.0580	0.4952	0.4913	0.0275 (0.0083)	0.0275 (0.0082)
0.7700	0.0510	0.0580	0.4929	0.5011	0.0275 (0.0082)	0.0274 (0.0085)
0.8650	0.0520	0.0610	0.4889	0.5003	0.0276 (0.0083)	0.0274 (0.0087)
0.9600	0.0570	0.0560	0.4843	0.4818	0.0278 (0.0084)	0.0278 (0.0084)
Mean	0.1919	0.1941	0.3862	0.3833	0.0327 (0.0098)	0.0328 (0.0098)

Table 2.2. *Coverage probability, p-value and mean(std) of the test statistic for grid $\mathcal{T}^{test}$, $n_0 = 6$; cpu time = 268.1657; cutoff value = 0.0428*

t_i	Coverage probability		p-value		Test statistic	
	RWA	GBS	RWA	GBS	RWA	GBS
0.1050	0.0640	0.0660	0.4688	0.4639	0.0282 (0.0086)	0.0284 (0.0085)
0.2000	0.0530	0.0500	0.5065	0.4918	0.0272 (0.0084)	0.0275 (0.0081)
0.2950	0.0450	0.0460	0.5066	0.4941	0.0272 (0.0083)	0.0275 (0.0083)
0.3900	0.0450	0.0500	0.4967	0.5078	0.0273 (0.0080)	0.0271 (0.0082)
0.4850	0.0500	0.0470	0.4929	0.5093	0.0274 (0.0082)	0.0270 (0.0081)
0.5800	0.0450	0.0500	0.5045	0.5072	0.0271 (0.0082)	0.0271 (0.0082)
0.6750	0.0470	0.0540	0.5080	0.5057	0.0271 (0.0081)	0.0271 (0.0082)
0.7700	0.0480	0.0540	0.5086	0.5016	0.0271 (0.0081)	0.0272 (0.0083)
0.8650	0.0520	0.0540	0.4999	0.5041	0.0274 (0.0084)	0.0272 (0.0084)
0.9600	0.0520	0.0520	0.5042	0.5059	0.0273 (0.0085)	0.0272 (0.0081)
Mean	0.0501	0.0523	0.4997	0.4991	0.0273 (0.0083)	0.0273 (0.0082)

Table 2.3. *Coverage probability, p-value and mean(std) of the test statistic for grid $\mathcal{T}^{test}$; $n_0 = 9$; cpu time = 603.1311; cutoff value = 0.0428*

We next perform some comparison tests between RWA and Bondesson's method. In each case, we mostly take the smallest values of n_0 and K that provide correct results. We first take again $A(t) = t$. The results are displayed in Table 2.6 for $n_0 = 6$ and $K = 40$ on grid $\mathcal{T}^{\text{gen}}$ used for generating the RWA points. The results are next provided on grid $\mathcal{T}^{\text{test}}$ in Table 2.7 for $n_0 = 9$ and $K = 40$.

t_i	Coverage probability			p-value		
	Bondesson	Rejection	Thinning	Bondesson	Rejection	Thinning
0.1050	0.0520	0.0500	1.0000	0.5071	0.5027	0.0000
0.2000	0.0560	0.0560	1.0000	0.4059	0.3924	0.0000
0.2950	0.0510	0.0450	1.0000	0.3776	0.3796	0.0000
0.3900	0.0580	0.0690	1.0000	0.3622	0.3563	0.0000
0.4850	0.0430	0.0530	1.0000	0.3592	0.3534	0.0000
0.5800	0.0440	0.0530	1.0000	0.3356	0.3279	0.0000
0.6750	0.0530	0.0550	1.0000	0.3215	0.3219	0.0000
0.7700	0.0550	0.0510	1.0000	0.3175	0.3134	0.0000
0.8650	0.0470	0.0420	1.0000	0.3054	0.3045	0.0000
0.9600	0.0600	0.0630	1.0000	0.2926	0.2887	0.0000
Mean	0.0519	0.0537	1.0000	0.3585	0.3541	0.0000

Table 2.4. *Coverage probability and p-value for grid $\mathcal{T}^{test}$; $K = 40$*

t_i	Bondesson	Rejection	Thinning
0.1050	0.0272 (0.0081)	0.0273 (0.0080)	0.6492 (0.0152)
0.2000	0.0294 (0.0070)	0.0297 (0.0069)	0.6754 (0.0150)
0.2950	0.0301 (0.0067)	0.0300 (0.0067)	0.6761 (0.0152)
0.3900	0.0305 (0.0068)	0.0307 (0.0069)	0.6756 (0.0151)
0.4850	0.0304 (0.0064)	0.0306 (0.0065)	0.6761 (0.0152)
0.5800	0.0310 (0.0063)	0.0313 (0.0063)	0.6756 (0.0148)
0.6750	0.0314 (0.0064)	0.0315 (0.0066)	0.6759 (0.0147)
0.7700	0.0315 (0.0062)	0.0316 (0.0060)	0.6761 (0.0148)
0.8650	0.0317 (0.0059)	0.0317 (0.0057)	0.6761 (0.0148)
0.9600	0.0322 (0.0061)	0.0324 (0.0063)	0.6757 (0.0148)
Mean	0.0305 (0.0066)	0.0307 (0.0066)	0.6732 (0.0150)

Table 2.5. *Mean(std) of the test statistic for grid $\mathcal{T}^{test}$; cutoff value = 0.0428; $K = 40$*

We next consider $A(t) = t^{0.8}$. The results are presented in Table 2.8 for grid $\mathcal{T}^{\text{gen}}$, $n_0 = 6$ and $K = 30$ and in Table 2.9 for grid $\mathcal{T}^{\text{test}}$, $n_0 = 9$ and $K = 30$.

We finally take $A(t) = t^{1.5}$. The results are presented in Table 2.10 for grid $\mathcal{T}^{\text{gen}}$, $n_0 = 6$ and $K = 100$ and in Table 2.11 for grid $\mathcal{T}^{\text{test}}$, $n_0 = 10$ and $K = 105$.

t_i	Coverage probability RWA	Coverage probability Bond.	p-value RWA	p-value Bond.	Test statistic RWA	Test statistic Bond.
0.1094	0.0440	0.0510	0.5118	0.4877	0.0270 (0.0082)	0.0276 (0.0081)
0.2031	0.0460	0.0490	0.4875	0.4960	0.0276 (0.0081)	0.0273 (0.0082)
0.2969	0.0410	0.0550	0.5191	0.4937	0.0268 (0.0080)	0.0275 (0.0083)
0.3906	0.0420	0.0520	0.5018	0.4938	0.0271 (0.0078)	0.0275 (0.0084)
0.4844	0.0500	0.0540	0.5123	0.4999	0.0270 (0.0082)	0.0273 (0.0084)
0.5781	0.0500	0.0540	0.5140	0.4985	0.0269 (0.0082)	0.0274 (0.0084)
0.6719	0.0520	0.0440	0.5039	0.5043	0.0272 (0.0082)	0.0272 (0.0083)
0.7656	0.0510	0.0520	0.5112	0.5039	0.0271 (0.0084)	0.0272 (0.0081)
0.8594	0.0570	0.0530	0.5032	0.4984	0.0273 (0.0086)	0.0273 (0.0083)
0.9531	0.0520	0.0500	0.5088	0.5038	0.0271 (0.0083)	0.0273 (0.0084)
Mean	0.0485	0.0514	0.5074	0.4980	0.0271 (0.0082)	0.0274 (0.0083)

Table 2.6. *Coverage probability, p-value and mean(std) of the test statistic for grid $\mathcal{T}^{gen}$; $n_0 = 6$; $K = 40$; Bondesson cpu time = 107.5939; RWA cpu time = 101.8219; cutoff value = 0.0428; $A(t) = t$*

t_i	Coverage probability RWA	Coverage probability Bond.	p-value RWA	p-value Bond.	Test statistic RWA	Test statistic Bond.
0.1050	0.0610	0.0520	0.4724	0.4882	0.0282 (0.0086)	0.0276 (0.0081)
0.2000	0.0600	0.0580	0.4695	0.4899	0.0281 (0.0083)	0.0276 (0.0083)
0.2950	0.0610	0.0520	0.4890	0.4913	0.0277 (0.0084)	0.0275 (0.0083)
0.3900	0.0570	0.0430	0.4866	0.4949	0.0277 (0.0085)	0.0274 (0.0082)
0.4850	0.0540	0.0530	0.4994	0.4910	0.0274 (0.0084)	0.0276 (0.0083)
0.5800	0.0580	0.0520	0.5026	0.4942	0.0273 (0.0083)	0.0275 (0.0083)
0.6750	0.0410	0.0440	0.5017	0.4970	0.0272 (0.0082)	0.0273 (0.0080)
0.7700	0.0480	0.0580	0.4918	0.5007	0.0275 (0.0081)	0.0274 (0.0085)
0.8650	0.0460	0.0600	0.4894	0.4969	0.0275 (0.0080)	0.0276 (0.0086)
0.9600	0.0510	0.0600	0.4825	0.4926	0.0277 (0.0080)	0.0275 (0.0084)
Mean	0.0537	0.0532	0.4885	0.4937	0.0276 (0.0083)	0.0275 (0.0083)

Table 2.7. *Coverage probability, p-value and mean(std) of the test statistic for grid $\mathcal{T}^{test}$; $n_0 = 9$; $K = 40$; Bondesson cpu time = 133.0221; RWA cpu time = 285.7626; cutoff value = 0.0428; $A(t) = t$*

t_i	Coverage probability		p-value		Test statistic	
	RWA	Bond.	RWA	Bond.	RWA	Bond.
0.1094	0.0540	0.0510	0.5061	0.4923	0.0271 (0.0081)	0.0275 (0.0084)
0.2031	0.0490	0.0520	0.5054	0.5004	0.0272 (0.0083)	0.0274 (0.0084)
0.2969	0.0450	0.0520	0.5040	0.4972	0.0271 (0.0080)	0.0274 (0.0082)
0.3906	0.0530	0.0500	0.4928	0.5055	0.0275 (0.0083)	0.0272 (0.0082)
0.4844	0.0400	0.0440	0.4997	0.5002	0.0273 (0.0082)	0.0272 (0.0082)
0.5781	0.0590	0.0500	0.5040	0.4958	0.0273 (0.0084)	0.0274 (0.0081)
0.6719	0.0540	0.0470	0.5109	0.4982	0.0271 (0.0083)	0.0273 (0.0080)
0.7656	0.0540	0.0450	0.5107	0.4921	0.0271 (0.0084)	0.0274 (0.0080)
0.8594	0.0580	0.0540	0.5124	0.4912	0.0271 (0.0085)	0.0275 (0.0082)
0.9531	0.0540	0.0570	0.5163	0.4864	0.0270 (0.0085)	0.0277 (0.0084)
Mean	0.0520	0.0502	0.5062	0.4959	0.0272 (0.0083)	0.0274 (0.0082)

Table 2.8. *Coverage probability, p-value and mean(std) of the test statistic for grid $\mathcal{T}^{gen}$; $n_0 = 6$; $K = 30$; Bondesson cpu time = 92.3058; RWA cpu time = 87.5322; cutoff value = 0.0428; $A(t) = t^{0.8}$*

t_i	Coverage probability		p-value		Test statistic	
	RWA	Bond.	RWA	Bond.	RWA	Bond.
0.1050	0.0530	0.0470	0.4948	0.5029	0.0275 (0.0083)	0.0272 (0.0081)
0.2000	0.0410	0.0340	0.5101	0.5103	0.0270 (0.0081)	0.0269 (0.0079)
0.2950	0.0490	0.0460	0.5001	0.5056	0.0272 (0.0080)	0.0271 (0.0080)
0.3900	0.0420	0.0370	0.5065	0.5049	0.0270 (0.0078)	0.0270 (0.0078)
0.4850	0.0460	0.0440	0.5085	0.5025	0.0270 (0.0079)	0.0271 (0.0080)
0.5800	0.0490	0.0450	0.4936	0.5024	0.0274 (0.0080)	0.0272 (0.0082)
0.6750	0.0530	0.0540	0.5000	0.4928	0.0273 (0.0083)	0.0275 (0.0083)
0.7700	0.0630	0.0500	0.5016	0.5009	0.0274 (0.0085)	0.0273 (0.0084)
0.8650	0.0560	0.0490	0.5064	0.4931	0.0273 (0.0088)	0.0275 (0.0081)
0.9600	0.0590	0.0500	0.5013	0.4961	0.0274 (0.0086)	0.0274 (0.0083)
Mean	0.0511	0.0456	0.5023	0.5011	0.0273 (0.0082)	0.0272 (0.0081)

Table 2.9. *Coverage probability, p-value and mean(std) of the test statistic for grid $\mathcal{T}^{test}$; $n_0 = 9$; $K = 30$; Bondesson cpu time = 91.791; RWA cpu time = 278.2746; cutoff value = 0.0428; $A(t) = t^{0.8}$*

t_i	Coverage probability		p-value		Test statistic	
	RWA	Bond.	RWA	Bond.	RWA	Bond.
0.1094	0.0420	0.0530	0.4825	0.3434	0.0277 (0.0081)	0.0308 (0.0063)
0.2031	0.0480	0.0390	0.5039	0.5092	0.0272 (0.0083)	0.0270 (0.0081)
0.2969	0.0570	0.0490	0.4961	0.5004	0.0275 (0.0085)	0.0272 (0.0079)
0.3906	0.0510	0.0530	0.4975	0.4928	0.0274 (0.0083)	0.0274 (0.0080)
0.4844	0.0470	0.0360	0.5141	0.4961	0.0269 (0.0082)	0.0274 (0.0082)
0.5781	0.0640	0.0340	0.4944	0.5124	0.0276 (0.0086)	0.0269 (0.0079)
0.6719	0.0490	0.0250	0.5005	0.5062	0.0273 (0.0082)	0.0269 (0.0076)
0.7656	0.0550	0.0400	0.5030	0.4958	0.0272 (0.0083)	0.0273 (0.0079)
0.8594	0.0510	0.0490	0.5032	0.4870	0.0272 (0.0082)	0.0276 (0.0079)
0.9531	0.0480	0.0540	0.4949	0.4891	0.0275 (0.0085)	0.0275 (0.0081)
Mean	0.0512	0.0432	0.4990	0.4832	0.0274 (0.0083)	0.0276 (0.0078)

Table 2.10. *Coverage probability, p-value and mean(std) of the test statistic for grid $\mathcal{T}^{gen}$; $n_0 = 6$; $K = 100$; Bondesson cpu time = 183.1608; RWA cpu time = 89.685; cutoff value = 0.0428; $A(t) = t^{1.5}$*

t_i	Coverage probability		p-value		Test statistic	
	RWA	Bond.	RWA	Bond.	RWA	Bond.
0.1050	0.0580	0.0490	0.5068	0.3199	0.0272 (0.0084)	0.0315 (0.0065)
0.2000	0.0630	0.0480	0.4847	0.5046	0.0278 (0.0085)	0.0271 (0.0079)
0.2950	0.0660	0.0420	0.4951	0.5087	0.0277 (0.0089)	0.0270 (0.0081)
0.3900	0.0620	0.0450	0.4878	0.5018	0.0278 (0.0086)	0.0272 (0.0081)
0.4850	0.0610	0.0530	0.4940	0.5082	0.0276 (0.0086)	0.0271 (0.0083)
0.5800	0.0610	0.0460	0.4944	0.5105	0.0275 (0.0084)	0.0270 (0.0082)
0.6750	0.0550	0.0540	0.4932	0.5067	0.0275 (0.0083)	0.0271 (0.0081)
0.7700	0.0540	0.0440	0.4932	0.4958	0.0275 (0.0082)	0.0273 (0.0080)
0.8650	0.0520	0.0380	0.4897	0.5030	0.0275 (0.0082)	0.0271 (0.0080)
0.9600	0.0470	0.0470	0.4898	0.5117	0.0275 (0.0081)	0.0270 (0.0080)
Mean	0.0579	0.0466	0.4929	0.4871	0.0276 (0.0084)	0.0276 (0.0079)

Table 2.11. *Coverage probability, p-value and mean(std) of the test statistic for grid $\mathcal{T}^{test}$; $n_0 = 10$; $K = 105$; Bondesson cpu time = 191.0544; RWA cpu time = 395.0101; cutoff value = 0.0428; $A(t) = t^{1.5}$*

Based on these tables, we can see that the RWA method behaves well when tested at points of its generation grid and even better than Bondesson's method in the case $A(t) = t^{1.5}$ (look at the computation times). However, when tested at points out of its generation grid, RWA method requires a longer computation time than Bondesson's method to get a similar accuracy.

As a conclusion to this section, it seems that the simple RWA method is well adapted when the point is to sample trajectories of a gamma process at fixed points known in advance. In the other case, Bondesson's (or the Rejection) method performs better.

2.6. Hitting time (time-to-failure) and overshoot

2.6.1. *Introduction*

In this section, $(X_t)_{t\geq 0}$ is a gamma process $\mathcal{G}(A(t), 1)$, where $A : \mathbb{R}_+ \to \mathbb{R}_+$ is assumed to be continuous, non-decreasing and such that $A(0) = 0$ and $a(\mathrm{d}t)$ is the only measure on $(\mathbb{R}_+, \mathcal{B}(\mathbb{R}_+))$ such that $A(t) = a([0,t])$, for all $t \geq 0$. As a non-decreasing function, $A(t)$ is almost surely differentiable (everywhere differentiable except on a set which is at most countable, see [RUD 87]). We set $a(s) = A'(s)$ a.s. so that $a(\mathrm{d}s) = a(s)\,\mathrm{d}s$. Finally, the trajectories of $(X_t)_{t\geq 0}$ are almost surely continuous so that

$$\mathbb{P}(X_t < x) = \mathbb{P}(X_t \leq x)$$

for all $t > 0$ and all $x \geq 0$.

In case X_t represents the deterioration accumulated by a system on the time interval $[0,t]$, the system is typically considered as failed as soon the deterioration level is beyond a given failure threshold (say h). The time-to-failure of the system hence is the hitting time τ_h of the Borel set $[h, +\infty)$ by the process $(X_t)_{t\geq 0}$, with

$$\tau_h = \inf(t \geq 0 : X_t \geq h).$$

Apart from their interpretation as failure times, a lot of preventive maintenance policies are also based on hitting times, the reaching of a preventive threshold m (with $m < h$) triggering a preventive maintenance action, see, e.g. [VAN 09] for a lot of references on the subject. Hitting times

τ_h (or τ_m) are hence of major importance. As a gamma process is a pure jump process, a threshold h is crossed by a jump and the after-jump level X_{τ_h} is greater than h almost surely ($X_{\tau_h} > h$ a.s.). The exceedance $X_{\tau_h} - h$ is known as the *overshoot* of the gamma process. Note that in an applicative context, the failure may be all the more severe as the overshoot is higher. Also, even if $m < h$, both levels m and h may be crossed at the same time τ_m in case of a big jump at time τ_m, so that we may have $X_{\tau_m} = X_{\tau_h}$. Hence the interest of studying the after-jump level X_{τ_h} jointly with the jump time τ_h. We begin with providing the joint distribution of $\left(\tau_h, X_{\tau_h^-}, X_{\tau_h}\right)$, as well as a few marginal distributions of $\left(\tau_h, X_{\tau_h^-}, X_{\tau_h}\right)$, where $X_{\tau_h^-}$ stands for the left-side limit of $(X_t)_{t\geq 0}$ at time τ_h, namely for the level just before the jump. We next study aging properties of the hitting time τ_h and we end with comparing the different methods at our disposal for simulating τ_h.

Note that if $(Y_t)_{t\geq 0}$ is a gamma process $\mathcal{G}(A(t), b)$ with $b > 0$, then

$$\tilde{\tau}_h = \inf(t \geq 0 : Y_t \geq h) = \inf(t \geq 0 : X_t \geq bh)$$

where $(X_t)_{t\geq 0} = (bY_t)_{\geq 0}$ is a gamma process $\mathcal{G}(A(t), 1)$, see proposition 2.25. Hence $\tilde{\tau}_h = \tau_{bh}$ and it is enough to consider the case where $b = 1$.

2.6.2. *Theoretical results*

Let us first recall that the p.d.f., c.d.f. and survival function of $\mathcal{G}(\alpha, \beta)$ are denoted by $f_{\alpha,\beta}$, $F_{\alpha,\beta}$ and $\bar{F}_{\alpha,\beta}$, respectively.

THEOREM 2.37.– The joint p.d.f. of $\left(\tau_h, X_{\tau_h^-}, X_{\tau_h}\right)$ is given as

$$f_{\left(\tau_h, X_{\tau_h^-}, X_{\tau_h}\right)}(s, x, z)$$
$$= \mathbf{1}_{\mathbb{R}_+}(s)\, \mathbf{1}_{\{0<x<h\leq z\}} f_{A(s),1}(x) \frac{e^{-(z-x)}}{z-x} a(s)\, \mathrm{d}s\, \mathrm{d}x\, \mathrm{d}z. \qquad [2.21]$$

PROOF.– The proof is based on [BER 96, Prop. 2 page 76] (in the homogeneous case) (see also [ÇIN 11, theorem 7.18 pp. 372]). Let φ be any non-negative and measurable function. Based on the fact that the jumps are at most countable, we have:

$$\mathbb{E}\left(\varphi\left(\tau_h, X_{\tau_h^-}, X_{\tau_h}\right)\right) = \sum_{s\geq 0} \mathbb{E}\left(\varphi\left(s, X_{s-}, X_s\right) \mathbf{1}_{\{X_{s-}<h\}} \mathbf{1}_{\{X_s \geq h\}}\right)$$

$$= \sum_{s\geq 0} \mathbb{E}\left(\varphi\left(s, X_{s-}, X_{s-} + \Delta X_s\right) \mathbf{1}_{\{X_{s-}<h\}} \mathbf{1}_{\{X_{s-}+\Delta X_s \geq h\}}\right)$$

where $\Delta X_s = X_s - X_{s-}$. Due to the predictability of $(X_{s-})_{s\geq 0}$ (as a left-continuous process) and remembering theorem 2.28, the compensation formula [BER 96, page 7] applied to the jump process $(\Delta X_s)_{s\geq 0}$ provides:

$$\mathbb{E}\left(\varphi\left(\tau_h, X_{\tau_h^-}, X_{\tau_h}\right)\right)$$

$$= \int_{\mathbb{R}_+^2} \mathbb{E}\left(\varphi\left(s, X_{s-}, X_{s-} + y\right) \mathbf{1}_{\{X_{s-}<h\}} \mathbf{1}_{\{X_{s-}+y \geq h\}}\right) \frac{e^{-y}}{y} \mathrm{d}y \, a\left(\mathrm{d}s\right).$$

As $a(\mathrm{d}s) = a(s)\ \mathrm{d}s$ and $X_s = X_{s-}$ almost surely with respect of Lebesgue measure, we get:

$$\mathbb{E}\left(\varphi\left(\tau_h, X_{\tau_h^-}, X_{\tau_h}\right)\right)$$

$$= \int_{\mathbb{R}_+^2} \mathbb{E}\left(\varphi\left(s, X_s, X_s + y\right) \mathbf{1}_{\{X_s<h\}} \mathbf{1}_{\{X_s+y \geq h\}}\right) \frac{e^{-y}}{y} \mathrm{d}y \, a\left(s\right)\ \mathrm{d}s$$

$$= \int_{\mathbb{R}_+^3} \varphi\left(s, x, x + y\right) \mathbf{1}_{\{x<h\}} \mathbf{1}_{\{x+y \geq h\}} f_{A(s),1}\left(x\right) \mathrm{d}x \, \frac{e^{-y}}{y} \mathrm{d}y \, a\left(s\right)\ \mathrm{d}s$$

$$= \int_{\mathbb{R}_+^3} \varphi\left(s, x, z\right) \mathbf{1}_{\{x<h\leq z\}} f_{A(s),1}\left(x\right) \frac{e^{-(z-x)}}{z-x}\ a\left(s\right)\ \mathrm{d}s\, \mathrm{d}x\, \mathrm{d}z$$

setting $z = x + y$ in the last line. This provides [2.21]. □

COROLLARY 2.38.– Assume that $\lim_{t\to+\infty} A(t) = \infty$. Then:

1) the joint p.d.f. of $\left(X_{\tau_h^-}, X_{\tau_h}\right)$ is *independent* of the shape function $A(t)$ and is:

$$f_{\left(X_{\tau_h^-}, X_{\tau_h}\right)}(x, z) = \mathbf{1}_{\{0<x<h\leq z\}} \frac{e^{-(z-x)}}{z-x} U(x)$$

where

$$U(x) = \int_{\mathbb{R}_+} f_{t,1}(x)\ \mathrm{d}t \qquad [2.22]$$

is the *potential function*.

2) The p.d.f. and survival functions of X_{τ_h} are

$$f_{X_{\tau_h}}(z) = \mathbf{1}_{\{h \leq z\}} \int_0^h \frac{e^{-(z-x)}}{z-x} U(x) \, \mathrm{d}x,$$

$$\bar{F}_{X_{\tau_h}}(z) = \int_0^h \mathrm{Ei}(z-x) \; U(x) \; \mathrm{d}x \text{ for all } z \geq h,$$

where we recall that Ei is the exponential integral function, see [2.16].

3) The p.d.f. of τ_h is

$$f_{\tau_h}(s) = a(s) \left(\mathrm{Ei} * f_{A(s),1} \right)(h) \text{ for all } s \geq 0. \qquad [2.23]$$

PROOF.– Starting from [2.21], we have:

$$\begin{aligned} f_{\left(X_{\tau_h^-}, X_{\tau_h}\right)}(x, z) &= \int_{\mathbb{R}_+} f_{\left(\tau_h, X_{\tau_h^-}, X_{\tau_h}\right)}(s, x, z) \; \mathrm{d}s \\ &= \mathbf{1}_{\{0<x<h \leq z\}} \int_{\mathbb{R}_+} f_{A(s),1}(x) \; \frac{e^{-(z-x)}}{z-x} a(s) \, \mathrm{d}s \\ &= \mathbf{1}_{\{0<x<h \leq z\}} \frac{e^{-(z-x)}}{z-x} \int_{\mathbb{R}_+} f_{t,1}(x) \; \mathrm{d}t \end{aligned}$$

setting $t = A(s)$ and remembering that $A(0) = 0$ and $\lim_{t \to +\infty} A(t) = \infty$. This provides the first point. For the second point, we have:

$$f_{X_{\tau_h}}(z) = \int_{\mathbb{R}_+} f_{\left(X_{\tau_h^-}, X_{\tau_h}\right)}(x, z) \; \mathrm{d}x = \mathbf{1}_{\{h \leq z\}} \int_0^h \frac{e^{-(z-x)}}{z-x} U(x) \; \mathrm{d}x$$

and, for $z \geq h$:

$$\begin{aligned} \bar{F}_{X_{\tau_h}}(z) &= \int_z^{+\infty} f_{X_{\tau_h}}(u) \; \mathrm{d}u \\ &= \int_z^{+\infty} \left(\int_0^h \frac{e^{-(u-x)}}{u-x} U(x) \; \mathrm{d}x \right) \mathrm{d}u \\ &= \int_0^h \left(\int_z^{+\infty} \frac{e^{-(u-x)}}{u-x} \; \mathrm{d}u \right) U(x) \; \mathrm{d}x \\ &= \int_0^h \mathrm{Ei}(z-x) \; U(x) \; \mathrm{d}x \end{aligned}$$

due to Fubini's theorem. The third point is similar. □

REMARK 2.39.– Note that though the distribution of τ_h depends on the shape function $A(t)$, the joint distribution of the before/after jump locations $\left(X_{\tau_h^-}, X_{\tau_h}\right)$ is *independent* of $A(t)$. Up to our knowledge, this has not been noticed before.

Before going on with more classical results from the reliability literature [VAN 09] (see also [PAR 14]), we first provide the definition of some special functions to be used in the following proposition (for further details and properties of these functions, the reader can refer to [ABR 72] or to [GRA 07]).

DEFINITION 2.40.–

– *The di-gamma function ψ_0 is the logarithmic derivative of the gamma function:*

$$\forall x > 0, \quad \psi_0(x) = \frac{\Gamma'(x)}{\Gamma(x)}.$$

– *The lower incomplete gamma function $\gamma(\cdot,\cdot)$ is defined as follows, for any $a > 0$ and $x > 0$:*

$$\gamma(a,x) = \int_0^a u^{x-1} e^{-u}\, du.$$

– *The upper incomplete gamma function $\Gamma(\cdot,\cdot)$ is the complementary of $\gamma(\cdot,\cdot)$ to the gamma function:*

$$\Gamma(a,x) = \Gamma(a) - \gamma(a,x) = \int_a^\infty u^{x-1} e^{-u}\, du$$

for all $a, x > 0$.

– *Let $p, q \in \mathbb{N}^*$. The generalized hyper-geometric function ${}_pF_q$ of order (p,q) is defined as follows, for any $z > 0$ and any positive real numbers $a_1, \ldots, a_p, b_1, \ldots, b_q$:*

$${}_pF_q(a_1, \ldots, a_p; b_1, \ldots, b_q; z) = \sum_{k=0}^{\infty} \frac{(a_1)_k \cdots (a_p)_k}{(b_1)_k \cdots (b_q)_k} \frac{z^k}{k!},$$

where $(x)_n = \Gamma(x+n)/\Gamma(x)$ is the Pochammer symbol.

PROPOSITION 2.41.– With the previous notations, for all $h > 0$, we have:

1) The survival function of τ_h is:

$$\bar{F}_{\tau_h}(t) = \mathbb{P}(X_t < h) = F_{A(t),1}(h).$$

$$f_{\tau_h}(t) = a(t)\int_h^{+\infty} \{\log(u) - \psi_0(A(t))\} f_{A(t),1}(u)\, \mathrm{d}u \qquad [2.24]$$

$$= a(t)\left(\psi_0(A(t)) - \log(h)\right)\frac{\gamma(A(t),h)}{\Gamma(A(t))} + \frac{a(t)h^{A(t)}}{A(t)^2\Gamma(A(t))}$$
$$\times {}_2F_2(A(t), A(t); A(t)+1, A(t)+1; -h). \qquad [2.25]$$

2) The mean time to failure is:

$$\mathbb{E}(\tau_h) = \int_0^{\infty} F_{A(s),1}(h)\, \mathrm{d}s \qquad [2.26]$$

and $\mathbb{E}(\tau_h) = \frac{1}{a}\int_0^h U(x)\, \mathrm{d}x$ in the homogeneous case $A(t) = at$, where $U(t)$ is the potential function given by [2.22]. Also:

$$\mathbb{E}(\tau_h^2) = \int_0^{\infty} F_{A(\sqrt{s}),1}(h)\, \mathrm{d}s.$$

PROOF.– Based on the fact that X_t is non-decreasing and right continuous, we have

$$\mathbb{P}(\tau_h > t) = \mathbb{P}(X_t < h), \qquad [2.27]$$

which provides the first point.

For the second point, we start from

$$\bar{F}_{\tau_h}(t) = F_{A(t),1}(h) = \frac{\int_0^h x^{A(t)-1}e^{-x}\, \mathrm{d}x}{\Gamma(A(t))} = 1 - \frac{\int_h^{+\infty} x^{A(t)-1}e^{-x}\, \mathrm{d}x}{\Gamma(A(t))} \qquad [2.28]$$

and we first check that $\Gamma(t)$ is differentiable and compute its derivative. We write

$$\Gamma(t) = \int_0^{+\infty} g(t,x)\, \mathrm{d}x$$

with

$$g(t,x) = x^{t-1}e^{-x} = e^{(t-1)\log(x)}e^{-x}.$$

We have:

$$\frac{\partial}{\partial t} g(t,x) = \log(x)\, x^{t-1}e^{-x} \text{ for all } t > 0 \text{ and all } x > 0$$

and for $0 < t_1 < t < t_2$:

$$\left|\frac{\partial}{\partial t} g(t,x)\right| \leq h(x) = |\log(x)| \left(x^{t_1-1} + x^{t_2-1}\right) e^{-x}$$

with $\int_0^{+\infty} h(x)\, \mathrm{d}x < +\infty$. Based on Lebesgue's dominated convergence theorem, we may use differentiation under the integral sign, which provides the differentiability of $\Gamma(t)$ and

$$\Gamma'(t) = \int_0^{+\infty} \frac{\partial}{\partial t} g(t,x)\, \mathrm{d}x = \int_0^{+\infty} \log(x)\, x^{t-1}e^{-x}\, \mathrm{d}x.$$

Using similar arguments as for $\Gamma(t)$, we get:

$$\frac{\partial}{\partial t}\left(\int_h^{+\infty} x^{A(t)-1}e^{-x}\, \mathrm{d}x\right) = a(t)\int_h^{+\infty} \log(x)\, x^{A(t)-1}e^{-x}\, \mathrm{d}x$$

where $a(t) = A'(t)$ a.s.

Starting again from [2.28], we now have:

$$\begin{aligned} f_{T_h}(t) &= \frac{a(t)}{\Gamma(A(t))} \\ &\times \left(\int_h^{+\infty} \log(x)\, x^{A(t)-1}e^{-x}\, \mathrm{d}x - \int_h^{+\infty} x^{A(t)-1}e^{-x}\, \mathrm{d}x \times \frac{\Gamma'(A(t))}{\Gamma(A(t))}\right) \\ &= \frac{a(t)}{\Gamma(A(t))}\left(\int_h^{+\infty} \{\log(x) - \psi_0(A(t))\}\, x^{A(t)-1}e^{-x}\mathrm{d}x\right). \end{aligned} \quad [2.29]$$

For the second expression of f_{τ_h}, we start again from equation [2.29] that we write as follows:

$$f_{\tau_h}(t) = a(t)g(A(t)) \quad [2.30]$$

with

$$g(x) = \frac{1}{\Gamma(x)}\frac{\partial}{\partial x}\Gamma(x,h) - \frac{1}{\Gamma(x)}\psi_0(x)\Gamma(x,h). \quad [2.31]$$

From equation (6.5.12) in [ABR 72], we have

$$\Gamma(x,h) = \Gamma(x) - \gamma(x,h) = \Gamma(x) - \frac{h^x}{x}\,{}_1F_1\left(x;x+1;-h\right). \quad [2.32]$$

Moreover, the following identity holds:

$$\frac{\mathrm{d}}{\mathrm{d}x}\left(\frac{x}{x+k}\right) = \frac{1}{x}\left(\frac{(x)_k}{(x+1)_k} - \left(\frac{(x)_k}{(x+1)_k}\right)^2\right).$$

From the definition of the generalized hyper-geometric function ${}_1F_1$ of order $(1,1)$ and using the above equation, we have

$$\begin{aligned}\frac{\partial}{\partial x}{}_1F_1\left(x;x+1;-h\right) &= \frac{1}{x}\left[{}_1F_1\left(x;x+1;-h\right)\right.\\ &\left.-{}_2F_2\left(x,x;x+1,x+1;-h\right)\right]. \quad [2.33]\end{aligned}$$

From equations [2.32] and [2.33], we have

$$\begin{aligned}\frac{\partial}{\partial x}\Gamma(x,h) &= \Gamma(x)(\psi_0(x) - \log(h))\\ &+\log(h)\Gamma(x,h) + \frac{h^x}{x}{}_2F_2\left(x,x;x+1,x+1;-h\right).\end{aligned}$$

Replacing this last expression in equation [2.31], we finally have

$$g(x) = (\psi_0(x) - \log(h))\left(1 - \frac{\Gamma(x,h)}{\Gamma(x)}\right) + \frac{h^x}{x^2}{}_2F_2\left(x,x;x+1,x+1;-h\right),$$

which provides the result based on [2.30].

For the third point, we write

$$\mathbb{E}(\tau_h) = \int_0^{+\infty} \bar{F}_{\tau_h}(t)\,\mathrm{d}t = \int_0^{+\infty} F_{A(t),1}(h)\ \mathrm{d}t,$$

$$\mathbb{E}(\tau_h^2) = \int_0^{+\infty} \mathbb{P}(\tau_h^2 > t)\,\mathrm{d}t = \int_0^{+\infty} \bar{F}_{\tau_h}\left(\sqrt{t}\right)\mathrm{d}t$$

$$= \int_0^{+\infty} F_{A(\sqrt{t}),1}(h)\ \mathrm{d}t.$$

Applying Fubini's theorem to the expression of $\mathbb{E}(\tau_h)$ in the homogeneous case and setting $s = at$, we get:

$$\mathbb{E}(\tau_h) = \int_0^{+\infty}\left(\int_0^h f_{at,1}(x)\,\mathrm{d}x\right)\mathrm{d}t = \frac{1}{a}\int_0^h\left(\int_0^{+\infty} f_{s,1}(x)\,\mathrm{d}s\right)\mathrm{d}x$$

$$= \frac{1}{a}\int_0^h U(x)\mathrm{d}x.$$

□

COROLLARY 2.42.– For all $h > 0$, we have:

$$\mathbb{E}(X_{\tau_h}) = \int_0^{\infty} F_{s,1}(h)\ \mathrm{d}s = \int_0^h U(x)\mathrm{d}x$$

(independent of $A(t)$) and in the homogeneous case $A(t) = at$, we have: $\mathbb{E}(X_{\tau_h}) = a\,\mathbb{E}(\tau_h)$.

PROOF.– As the distribution of X_{τ_h} does not depend on $A(t)$ (see corollary 2.38), it is sufficient to consider the standard case $\mathcal{G}(t,1)$. Using the fact that $(X_t - t)_{t\geq 0}$ is a martingale that is uniformly integrable on each $[0, t_0]$ and that τ_h is a stopping time, we obtain that

$$\mathbb{E}\left(X_{\min(\tau_h, t_0)}\right) = \mathbb{E}(\min(\tau_h, t_0))$$

for all $t_0 \geq 0$ (see, e.g. [DOO 53] for more details on these notions). Letting $t_0 \to \infty$, we get by monotonous convergence that

$$\mathbb{E}(X_{\tau_h}) = \lim_{t_0\to\infty} \mathbb{E}\left(X_{\min(\tau_h, t_0)}\right) = \mathbb{E}(\tau_h),$$

which provides the first result due to [2.26]. The remaining is clear. □

EXAMPLE 2.43.– To better understand the behavior of τ_h and of X_{τ_h}, their p.d.f.s are plotted in Figure 2.3 in the case of a standard gamma process $\mathcal{G}(t,1)$

and $h = 2$. In this case, we have $\mathbb{E}(\tau_h) = \mathbb{E}(X_{\tau_h}) \simeq 2.50$, $\text{Var}(\tau_h) \simeq 1.94$, $\text{Var}(X_{\tau_h}) \simeq 0.45$ and $\mathbb{P}(X_{\tau_h} > 3) \simeq 0.15$. Based on the figure and on these results, we can see that X_{τ_h} may be largely beyond h (and it is not a rare event). Considering that $X_{\tau_h} \simeq h$ (namely, not taking the overshoot of the gamma process into account) as is sometimes done in reliability engineering may hence lead to a clear underestimation of the deterioration level at time τ_h and hence to some eventually hazardous situation.

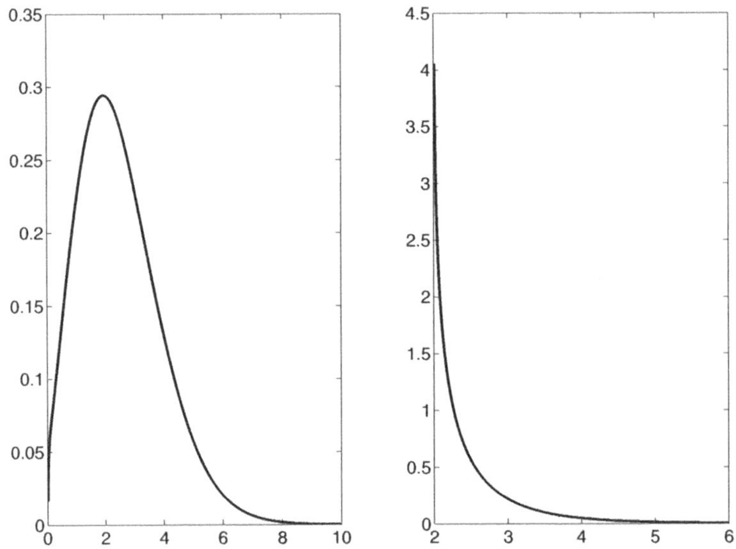

Figure 2.3. *p.d.f. of τ_h (left) and of X_{τ_h} (right) for a standard gamma process $\mathcal{G}(t, 1)$ and $h = 2$*

We next come to some *aging* properties of the hitting time τ_h, which indicates that a system with non-decreasing gamma deterioration becomes more prone to failures when it gets older. More details on these notions may be found in [SHA 07] for example. These aging properties justify the use of preventive actions to rejuvenate the system and prevent failures, when it is getting too old (or too deteriorated).

Let us consider a system with lifetime U and let U_t stand for the *remaining (residual) lifetime* of the system at time t (given that it has not failed at time

t). We recall that U_t is identically distributed as the conditional distribution of $U - t$ given that $U > t$. The survival function of U_t is

$$\bar{F}_{U_t}(u) = \mathbb{P}(U - t > u | U > t) = \frac{\bar{F}_U(t+u)}{\bar{F}_U(t)} \text{ for all } u \geq 0$$

and all t such that $\bar{F}_U(t) > 0$. Now, let us recall that a non-negative random variable U is said to have an increasing failure rate (IFR) as soon as $\frac{\bar{F}_U(t+u)}{\bar{F}_U(t)}$ is non-increasing in $t \geq 0$ (on the open support of $\bar{F}_U$), for each $u \geq 0$. Based on the previous interpretation, the IFR property means that

$$\bar{F}_{U_{t_1}}(u) \leq \bar{F}_{U_{t_2}}(u)$$

for all $t_1 \geq t_2$ and all $u \geq 0$. Equivalently, it means that the remaining lifetime of the system at time t stochastically decreases with respect to t in the sense of the usual stochastic order (see definition 2.7). If U admits a p.d.f. f_U with respect of Lebesgue measure, the IFR property is equivalent to the non-decreasingness of the failure rate function

$$h_U(u) = \begin{cases} \frac{f_U(u)}{\bar{F}_U(u)} \text{ if } \bar{F}_U(u) > 0, \\ \infty \text{ elsewhere,} \end{cases}$$

hence the name.

In the same way, a non-negative random variable U is said to be new better than used (NBU) if, for all $t \geq 0$, the remaining lifetime of the system at time t is stochastically smaller than the lifetime of a new system, namely if

$$\bar{F}_{U_t}(u) = \frac{\bar{F}_U(t+u)}{\bar{F}_U(t)} \leq \bar{F}_U(u)$$

for all $u \geq 0$ and all t in the open support of $\bar{F}_U$. We recall that the IFR property implies the NBU property.

We are now ready to give aging properties of the hitting time τ_h of a gamma process. An alternate proof of the following proposition may be found in [PAR 14].

PROPOSITION 2.44.– Let $h > 0$.

1) If $A(t)$ is superadditive (namely $A(t+u) \geq A(u) + A(t)$ for all $u, t \geq 0$), then τ_h is NBU;

2) If $A(t)$ is convex, then τ_h is IFR.

PROOF.– Based on the fact that $X_t \geq 0$, $X_{t+u} - X_t \geq 0$ and the independent increments, we have:

$$\begin{aligned} \bar{F}_{\tau_h}(t+u) &= \mathbb{P}(X_t + (X_{t+u} - X_t) \leq h) \\ &\leq \mathbb{P}(\{X_t \leq h\} \cap \{X_{t+u} - X_t \leq h\}) \\ &\leq \mathbb{P}(X_t \leq h)\ \mathbb{P}(X_{t+u} - X_t \leq h) \end{aligned} \qquad [2.34]$$

where $X_{t+u} - X_t$ is $\mathcal{G}(A(t+u) - A(t), 1)$ distributed. Assuming $A(t)$ to be superadditive, we have $A(t+u) - A(t) \geq A(u)$. Based on proposition 2.8, we know that $X_{t+u} - X_t \succ_{sto} X_u$ and consequently that

$$\mathbb{P}(X_{t+u} - X_t \leq h) \leq \mathbb{P}(X_u \leq h).$$

Starting again from [2.34], we obtain that $\bar{F}_{\tau_h}(t+u) \leq \bar{F}_{\tau_h}(t)\bar{F}_{\tau_h}(u)$. Hence τ_h is NBU, which provides the first point.

As for the second point, let us write

$$\bar{F}_{\tau_h}(t+u) = \mathbb{P}(X_{t+u} \leq h) = \mathbb{E}\left(\mathbf{1}_{\{X_t \leq h - (X_{t+u} - X_t)\}}\right).$$

Conditioning by $X_{t+u} - X_t$ and using the independent increments of $(X_t)_{t\geq 0}$, we obtain:

$$\bar{F}_{\tau_h}(t+u) = \mathbb{E}\left(F_{A(t),1}(h - (X_{t+u} - X_t))\right)$$

and

$$\frac{\bar{F}_{\tau_h}(t+u)}{\bar{F}_{\tau_h}(t)} = \frac{\mathbb{E}\left(F_{A(t),1}(h - (X_{t+u} - X_t))\right)}{F_{A(t),1}(h)}.$$

Let $t_1 < t_2$. We then have $A(t_1) \leq A(t_2)$ and $X_{t_1} \prec_{rhr} X_{t_2}$ (see proposition 2.8). We derive that

$$\frac{F_{A(t_1),1}(h - (X_{t_1+u} - X_{t_1}))}{F_{A(t_1),1}(h)} \geq \frac{F_{A(t_2),1}(h - (X_{t_1+u} - X_{t_1}))}{F_{A(t_2),1}(h)}$$

and consequently that

$$\frac{\bar{F}_{\tau_h}(t_1+u)}{\bar{F}_{\tau_h}(t_1)} = \frac{\mathbb{E}\left(F_{A(t_1),1}\left(h-(X_{t_1+u}-X_{t_1})\right)\right)}{F_{A(t_1),1}(h)}$$
$$\geq \frac{\mathbb{E}\left(F_{A(t_2),1}\left(h-(X_{t_1+u}-X_{t_1})\right)\right)}{F_{A(t_2),1}(h)}. \qquad [2.35]$$

Assuming $A(t)$ to be convex, we have $A(t_1+u)-A(t_1) \leq A(t_2+u)-A(t_2)$ so that $X_{t_1+u}-X_{t_1} \prec_{sto} X_{t_2+u}-X_{t_2}$. As $\varphi(x) = F_{A(t_2),1}(h-x)$ is non-increasing, we derive (see [SHA 07]) that

$$\mathbb{E}\left(F_{A(t_2),1}\left(h-(X_{t_1+u}-X_{t_1})\right)\right) \geq \mathbb{E}\left(F_{A(t_2),1}\left(h-(X_{t_2+u}-X_{t_2})\right)\right)$$
$$= \bar{F}_{\tau_h}(t_2+u).$$

Starting again from [2.35], we now have

$$\frac{\bar{F}_{\tau_h}(t_1+u)}{\bar{F}_{\tau_h}(t_1)} \geq \frac{\bar{F}_{\tau_h}(t_2+u)}{\bar{F}_{\tau_h}(t_2)}$$

and τ_h is IFR. □

2.6.3. *Simulation*

We here simulate trajectories of a gamma process on a time interval $[0, T]$, where T is chosen such that $\mathbb{P}(\tau_h > T)$ is very small. This allows us to observe τ_h for nearly almost all simulated trajectories. An alternate method would be (1) to generate step by step the trajectory by the RWA method, (2) to test whether h has been exceeded at each step and (3) to stop as soon as h has been exceeded. However, this entails a lot of tests and cannot be implemented in a matrix way. This consequently entails longer computing times.

We take $h = 2$ and several choices for $A(t)$. We simulate 1,000 independent data sets composed each of 1,000 independent observations of τ_h. For each data set of size 1,000, we test the adequation of the observations to the distribution of τ_h with the help of a Kolmogorov–Smirnov test at the significance level $\alpha = 0.05$. For each of these 1,000 tests, we compute the p-value and the test statistic. In the following tables, we provide the proportion of times the null assumption is rejected among the 1,000 tests (the coverage probability), which should be around 0.05, as well as the mean of the 1,000 computed p-values and both the mean and standard deviation of the 1,000 test statistics. The c.p.u. times are also given.

As a first case, we consider $A(t) = t$. With $h = 2$, this provides $\mathbb{E}(\tau_h) = 2.4961$ and $\text{Var}(\tau_h) = 1.9375$. The trajectories are simulated on $[0, T]$ with $T = 15$, so that $\mathbb{P}(\tau_h > T) = 3.87 \times 10^{-9}$ and the probability that all the $10^3 \times 10^3 = 10^6$ trajectories reach h is 0.9961. The results are provided in Table 2.12.

As a second case, we take $A(t) = t^{0.8}$. With $h = 2$, we get $\mathbb{E}(\tau_h) = 3.2896$ and $\text{Var}(\tau_h) = 5.0364$. The trajectories are simulated on $[0, T]$ with $T = 30$, so that $\mathbb{P}(\tau_h > T) = 2.59 \times 10^{-9}$ and the probability that all the 10^6 trajectories reach h is 0.9974. The results are provided in Table 2.13.

		Probability of rejecting $\mathcal{H}_0$		p-value		Test statistic	
n_0	K	RWA	Bond.	RWA	Bond.	RWA	Bond.
8	5	0.1720	0.0530	0.2854	0.4779	0.0342(0.0097)	0.0279(0.0085)
9	6	0.0930	0.0440	0.3930	0.5027	0.0303(0.0084)	0.0272(0.0080)
10	7	0.0520	0.0510	0.4569	0.4864	0.0284(0.0081)	0.0277(0.0083)
11	8	0.0420	0.0560	0.4964	0.4894	0.0273(0.0079)	0.0276(0.0082)

c.p.u. times				
(n_0, K)	$(8, 5)$	$(9, 6)$	$(10, 7)$	$(11, 8)$
(RWA, Bond.)	(150, 58)	(242, 53)	(380, 63)	(645, 70)

Table 2.12. *Coverage probability, p-value, mean(std) of the test statistic and c.p.u. times; $h = 2$; cutoff value = 0.0428; $A(t) = t$*

		Probability of rejecting $\mathcal{H}_0$		p-value		Test statistic	
n_0	K	RWA	Bond.	RWA	Bond.	RWA	Bond.
8	5	0.1120	0.0450	0.3527	0.4682	0.0316(0.0087)	0.0280(0.0080)
9	6	0.0640	0.0410	0.4662	0.5047	0.0284(0.0087)	0.0272(0.0082)
10	7	0.0660	0.0480	0.4687	0.4949	0.0282(0.0086)	0.0274(0.0083)
11	8	0.0510	0.0470	0.5020	0.5140	0.0272(0.0081)	0.0270(0.0084)

c.p.u. times				
(n_0, K)	$(8, 5)$	$(9, 6)$	$(10, 7)$	$(11, 8)$
(RWA, Bond.)	(178, 59)	(293, 63)	(417, 68)	(694, 78)

Table 2.13. *Coverage probability, p-value and mean(std) of the test statistic; $h = 2$; cutoff value = 0.0428; $A(t) = t^{0.8}$*

As a third and last case, we take $A(t) = t^{1.5}$. With $h = 2$, we get $\mathbb{E}(\tau_h) = 1.7721$ and $\mathrm{Var}(\tau_h) = 0.4782$. The trajectories are simulated on $[0, T]$ with $T = 6$, so that $\mathbb{P}(\tau_h > T) = 7.20 \times 10^{-9}$ and the probability that all the 10^6 trajectories reach h is 0.9928. The results are provided in Table 2.14.

		Probability of rejecting $\mathcal{H}_0$		p-value		Test statistic	
n_0	K	RWA	Bond.	RWA	Bond.	RWA	Bond.
8	5	0.1220	0.0660	0.3360	0.4938	0.0322(0.0089)	0.0276(0.0087)
9	7	0.0870	0.0460	0.4311	0.5046	0.0294(0.0089)	0.0271(0.0080)
10	9	0.0670	0.0440	0.4872	0.5188	0.0277(0.0083)	0.0268(0.0080)
11	11	0.0470	0.0560	0.4831	0.4964	0.0278(0.0083)	0.0275(0.0085)

c.p.u. times				
(n_0, K)	$(8, 5)$	$(9, 7)$	$(10, 9)$	$(11, 11)$
(RWA, Bond.)	(154, 64)	(231, 87)	(369 ,108)	(631 ,125)

Table 2.14. *Coverage probability, p-value and mean(std) of the test statistic; $h = 2$; cutoff value = 0.0428; $A(t) = t^{1.5}$*

Based on these tables, the RWA method requires much longer computing times than Bondesson's (or the Rejection) method to get a similar accuracy for the hitting time τ_h. Though the results are not provided here, the conclusion is similar for the levels reached just before ($X_{\tau_{h-}}$) or just after (X_{τ_h}) the level h is crossed.

In the case where it is desirable to observe hitting times or the corresponding deterioration levels, we consequently suggest to use Bondesson's (or the Rejection) method for simulating a gamma process.

2.7. Statistical inference in parametric models

In this section, we consider the degradation model based on a gamma process X that initiates at time t_0 with initial degradation level equal to x_0:

$$\forall t \geq 0, \quad X_t = (x_0 + \tilde{X}_{t-t_0})\mathbf{1}_{t \geq t_0}, \qquad [2.36]$$

where $\tilde{X}$ is a standard gamma process (i.e. starting from $\tilde{X}_0 = 0$ at time $t = 0$) with shape function $A(\cdot)$ and rate parameter b as defined previously.

The expectation is then equal to:

$$\mathbb{E}[X_t] = \begin{cases} 0 & \text{if } t < t_0 \\ x_0 + \frac{A(t-t_0)}{b} & \text{otherwise} \end{cases}$$

and the variance is equal to:

$$\text{Var}[X_t] = \begin{cases} 0 & \text{if } t < t_0 \\ \frac{A(t-t_0)}{b^2} & \text{otherwise} \end{cases}$$

We can notice that, in particular, we have $\mathbb{E}[X_{t_0}] = x_0$ and $\text{Var}[X_{t_0}] = 0$ since $A(0) = 0$.

For such a model, we provide estimators of the parameters and some of their asymptotic properties. The case where $t_0 = 0$ and/or $x_0 = 0$ is detailed. Two kinds of data can be distinguished: degradation measurements and failure times. We will consider three different cases: only degradation data are available, or only failure data are available or both kinds of data are available.

Here, we assume that the shape function depends on some unknown parameters $\theta \in \Theta \subseteq \mathbb{R}^p$. Thus, we will denote by $A(\cdot;\theta)$ the shape function in this framework. The two principal choices for $A(\cdot;\theta)$ are the following ones: (1) the power shape function: $A(t;\theta) = at^\gamma$ and thus $\theta = (a,\gamma)$ with $a, \gamma > 0$; (2) the exponential shape function: $A(t;\theta) = e^{at} - 1$ and thus $\theta = a > 0$. In the former case, when $\gamma = 1$ (linear shape function), it turns to be the homogeneous gamma process. In the later case, for a small value of a and at the beginning of the degradation, the "behavior" of the process is close to that of a homogeneous gamma process.

In some cases, the critical level h is an unknown parameter that should be estimated too. It follows that we will consider the problem of estimating either (t_0, x_0, θ, b) or (t_0, x_0, θ, b, h).

We now introduce some notations. We assume that we observe n independent and identically distributed copies of the stochastic process X described by equation [2.36], denoted by $(X_t^{(1)})_{t\geq 0}, \ldots, (X_t^{(n)})_{t\geq 0}$. The i-th sample path is observed m_i times at the instants $t_0 < t_{i,1} < \cdots < t_{i,m_i}$.

For any $i \in \{1, \ldots, n\}$ and any $j \in \{1, \ldots, m_i\}$, we set:

– $X_{i,j} = X_{t_{i,j}}^{(i)}$;

– $s_{i,j} = t_{i,j} - t_{i,j-1}$ with $t_{i,0} = t_0$;

– $Y_{i,j} = X_{i,j} - X_{i,j-1}$ with $X_{i,0} = x_0$;

– $\tau_h^{(i)} = \inf\{t \geq 0 \ ; \ X_t^{(i)} \geq c\}$;

– $A_{i,j}(\theta) = A(t_{i,j}; \theta)$;

– $\Delta A_{i,j}(\theta) = A(t_{i,j}; \theta) - A(t_{i,j-1}; \theta)$;

– $\mathbf{D}_{obs} = \bigcup_{i=1}^{n} \bigcup_{j=1}^{m_i} \{(t_{i,j}, X_{i,j})\}$ (set of all the observed degradation data).

Clearly, for any $i \in \{1, \ldots, n\}$ and any $j \in \{1, \ldots, m_i\}$, $Y_{i,j}$ is gamma distributed with shape parameter $\Delta A_{i,j}(\theta)$ and rate parameter b. For a linear shape function (homogeneous gamma process), $Y_{i,j}$ is simply gamma distributed with shape parameter $as_{i,j}$ and rate parameter b. In addition, when the degradation processes are observed the same number m of times at regular instants (called, hereafter, balanced case), all increments are identically distributed. Indeed, we have $m_i = m$ for any $i \in \{1, \ldots, n\}$ and $s_{i,j} = \delta$ for any $i \in \{1, \ldots, n\}$ and for any $j \in \{1, \ldots, m\}$. Thus, $Y_{1,1}, \ldots, Y_{1,m}, \ldots, Y_{n,1}, \ldots, Y_{n,m}$ are nm independent and identically distributed gamma distributed random variables with shape function $a\delta$ and rate parameter b.

2.7.1. *Estimation based only on degradation data*

Here, we will estimate parameters of the degradation model using only degradation data. Two different approaches are considered: the MLE and the moments method estimator (MME). In this case, the parameter h cannot be estimated (this is obvious since no information on failure times is available). We will consider two different situations. The first situation corresponds to the general case where (t_0, x_0, θ, b) are unknown, while the second situation corresponds to the simpler case where only (θ, b) are unknown. For the sake of simplicity and without loss of generality, in this situation, we assume that $t_0 = 0$ and $x_0 = 0$. For all these situations, we first consider the general case and next detail the balanced case.

2.7.1.1. *Maximum likelihood estimator*

For the different models described above, we provide the likelihood function and the equations satisfied by the MLE. Moreover, in some cases, we also give asymptotic results:

– *Case 1*: the estimation of t_0 and x_0 is based on the first increments, as we did for the Wiener process (see section 1.6.1). For the parameters θ and b, all the observations are used. The likelihood function is given by

$$L_1(t_0, x_0, \theta, b|\mathbf{D}_{obs})$$
$$= \prod_{i=1}^{n} \frac{b^{A(t_{i,1};\theta)-A(t_0;\theta)}}{\Gamma(A(t_{i,1};\theta) - A(t_0;\theta))}(X_{i,1} - x_0)^{A(t_{i,1};\theta)-A(t_0;\theta)-1}e^{-b(X_{i,1}-x_0)}$$
$$\times \prod_{i=1}^{n}\prod_{j=2}^{m_i} \frac{b^{\Delta A_{i,j}(\theta)}}{\Gamma(\Delta A_{i,j}(\theta))} Y_{i,j}^{\Delta A_{i,j}(\theta)-1} e^{-bY_{i,j}}. \quad [2.37]$$

The MLE can be computed numerically by maximizing the log-likelihood function:

$$\ell_1(t_0, x_0, \theta, b|\mathbf{D}_{obs}) = \log L_1(t_0, x_0, \theta, b|\mathbf{D}_{obs}).$$

Let us consider now the special case of a homogeneous gamma process: $A(t;\theta) = at$ for any $t \geq 0$. For such a model, the expression of the likelihood function can be a little simplified:

$$L_1(t_0, x_0, a, b|\mathbf{D}_{obs}) = \prod_{i=1}^{n} \frac{b^{a(t_{i,1}-t_0)}}{\Gamma(a(t_{i,1} - t_0))}$$
$$\times (X_{i,1} - x_0)^{a(t_{i,1}-t_0)-1}e^{-b(X_{i,1}-x_0)} \prod_{i=1}^{n}\prod_{j=2}^{m_i} \frac{b^{as_{i,j}}}{\Gamma(as_{i,j})} Y_{i,j}^{as_{i,j}-1} e^{-bY_{i,j}}$$
$$[2.38]$$

and thus the log-likelihood function is given by

$$\ell_1(t_0, x_0, a, b|\mathbf{D}_{obs})$$
$$= \sum_{i=1}^{n} \{a(t_{i,1} - t_0)\log b - \log\Gamma(a(t_{i,1} - t_0)) + (a(t_{i,1} - t_0) - 1)$$
$$\log(X_{i,1} - x_0) - b(X_{i,1} - x_0)\}$$
$$+\sum_{i=1}^{n}\sum_{j=2}^{m_i} \{as_{i,j}\log b - \log\Gamma(as_{i,j}) + (as_{i,j} - 1)\log Y_{i,j} - bY_{i,j}\}.$$

By differentiating the log-likelihood function ℓ_1 with respect to the four variables, the MLE $(\widehat{t_0}, \widehat{x_0}, \widehat{a}, \widehat{b})$ of (t_0, x_0, a, b) is the solution of the following set of four equations:

$$\left\{\begin{array}{ll}
\displaystyle\sum_{i=1}^{n} \psi_0(\widehat{a}(t_{i,1} - \widehat{t_0})) = n \log \widehat{b} + \sum_{i=1}^{n} \log(X_{i,1} - \widehat{x}_0) & [2.39a] \\
\displaystyle\widehat{b} = \sum_{i=1}^{n} \frac{\widehat{a}(t_{i,1} - \widehat{t_0}) - 1}{X_{i,1} - \widehat{x}_0} & [2.39b] \\
\displaystyle\sum_{i=1}^{n}(t_{i,1} - \widehat{t_0})\psi_0(\widehat{a}(t_{i,1} - \widehat{t_0})) + \sum_{i=1}^{n}\sum_{j=2}^{m_i} s_{i,j}\psi_0(\widehat{a}s_{i,j}) & \\
\displaystyle= \sum_{i=1}^{n}\sum_{j=2}^{m_i} s_{i,j} \log Y_{i,j} & \\
\displaystyle+ \log \widehat{b} \left(\sum_{i=1}^{n}(t_{i,1} - \widehat{t_0}) + \sum_{i=1}^{n}\sum_{j=2}^{m_i} s_{i,j} \right) + \sum_{i=1}^{n}(t_{i,1} - \widehat{t_0}) \log(X_{i,1} - \widehat{x}_0) & \\
 & [2.39c] \\
\displaystyle\widehat{b} = \widehat{a} \frac{\sum_{i=1}^{n}(t_{i,1} - \widehat{t_0}) + \sum_{i=1}^{n}\sum_{j=2}^{m_i} s_{i,j}}{\sum_{i=1}^{n}(X_{i,1} - \widehat{x}_0) + \sum_{i=1}^{n}\sum_{j=2}^{m_i} Y_{i,j}} & [2.39d]
\end{array}\right.$$

This offers two ways to determine the MLE numerically, either by numerically maximizing the log-likelihood function ℓ_1 or by solving the above set of equations. From equations [2.39b] and [2.39d], $\widehat{a}$ can be expressed with respect of $\widehat{t_0}$ and $\widehat{x}_0$ as follows:

$$\widehat{a} \left(\sum_{i=1}^{n} \frac{t_{i,1} - \widehat{t_0}}{X_{i,1} - \widehat{x}_0} - \frac{\sum_{i=1}^{n}(t_{i,1} - \widehat{t_0}) + \sum_{i=1}^{n}\sum_{j=2}^{m_i} s_{i,j}}{\sum_{i=1}^{n}(X_{i,1} - \widehat{x}_0) + \sum_{i=1}^{n}\sum_{j=2}^{m_i} Y_{i,j}} \right) = \sum_{i=1}^{n} \frac{1}{X_{i,1} - \widehat{x}_0}. \qquad [2.40]$$

Next, from equations [2.40] and [2.39b] (or equivalently [2.39d]), we can also express $\widehat{b}$ as a function of $\widehat{t_0}$ and $\widehat{x}_0$. Finally, replacing these expressions for $\widehat{a}$ and $\widehat{b}$ in equations [2.39a] and [2.39c], we obtain two equations depending only on $\widehat{t_0}$ and $\widehat{x}_0$. However, these equations cannot be simplified anymore and the solution cannot be easily determined. Thus, in practice, it is often easier to numerically maximize the log-likelihood function ℓ_1.

– *Case 2*: we now consider the situation where $t_0 = 0$ and $x_0 = 0$. The MLE $(\widehat{\theta}, \widehat{b})$ of (θ, b) can be obtained by maximizing the following likelihood function:

$$L_1(\theta, b|\mathbf{D}_{obs}) = \prod_{i=1}^{n}\prod_{j=1}^{m_i} \frac{b^{\Delta A_{i,j}(\theta)}}{\Gamma(\Delta A_{i,j}(\theta))} Y_{i,j}^{\Delta A_{i,j}(\theta)-1} e^{-bY_{i,j}} \qquad [2.41]$$

and the log-likelihood function is given as (up to an additive constant):

$$\ell_1(\theta, b|\mathbf{D}_{obs}) = \sum_{i=1}^{n}\sum_{j=1}^{m_i} (\Delta A_{i,j}(\theta)\log(b) - \log\Gamma(\Delta A_{i,j}(\theta)) + \Delta A_{i,j}(\theta)\log(Y_{i,j}) - bY_{i,j}). \qquad [2.42]$$

When the shape function is linear, $(\widehat{a}, \widehat{b})$ satisfies the following equations:

$$\begin{cases} \widehat{b} = \widehat{a}\dfrac{\sum_{i=1}^{n}\sum_{j=1}^{m_i} s_{i,j}}{\sum_{i=1}^{n}\sum_{j=1}^{m_i} Y_{i,j}} & [2.43a] \\ \sum_{i=1}^{n}\sum_{j=1}^{m_i} s_{i,j}\psi_0(\widehat{a}s_{i,j}) = \sum_{i=1}^{n}\sum_{j=1}^{m_i} s_{i,j}\log(\widehat{b}Y_{i,j}) & [2.43b] \end{cases}$$

Substituting the expression of $\widehat{b}$ from equation [2.43a] into equation [2.43b], we get

$$\sum_{i=1}^{n}\sum_{j=1}^{m_i} s_{i,j}\left(\psi_0(\widehat{a}s_{i,j}) - \log\widehat{a}\right) = \left(\sum_{i=1}^{n}\sum_{j=1}^{m_i} s_{i,j}\right)\log\left(\frac{\sum_{i=1}^{n}\sum_{j=1}^{m_i} s_{i,j}}{\sum_{i=1}^{n}\sum_{j=1}^{m_i} Y_{i,j}}\right) + \sum_{i=1}^{n}\sum_{j=1}^{m_i} s_{i,j}\log Y_{i,j}. \qquad [2.44]$$

Hence, the estimator $\widehat{a}$ can be numerically computed by solving this last equation and $\widehat{b}$ can next be computed using equation [2.43a].

Now, let us consider the balanced case where $m_i = m$ for any $i \in \{1, \ldots, n\}$ and $s_{i,j} = \delta$ for any $i \in \{1, \ldots, n\}$ and any $j \in \{1, \ldots, m\}$.

As noticed earlier, this situation is the basic one in statistical inference since we deal with i.i.d. random variables. In such a case, equation [2.44] can be written in terms of the empirical arithmetic and geometric averages of the $Y_{i,j}$'s as:

$$\begin{cases} \psi_0(\widehat{a}\delta) - \log(\widehat{a}\delta) = \overline{\log Y} - \log \overline{Y} & [2.45a] \\ \widehat{b} = \dfrac{\widehat{a}\delta}{\overline{Y}}, & [2.45b] \end{cases}$$

where

$$\overline{Y} = \frac{1}{mn} \sum_{i=1}^{n} \sum_{j=1}^{m} Y_{i,j} \quad \text{and} \quad \overline{\log Y} = \frac{1}{mn} \sum_{i=1}^{n} \sum_{j=1}^{m} \log Y_{i,j}.$$

Moreover, the MLE is known to be asymptotically normal (see [KOT 06], for instance). More specifically, we have:

$$\sqrt{nm} \begin{pmatrix} \widehat{a} - a \\ \widehat{b} - b \end{pmatrix} \xrightarrow[nm \to \infty]{d} \mathcal{N}(0, \Sigma)$$

where

$$\Sigma = \frac{1}{a\psi_1(a) - 1} \begin{pmatrix} \dfrac{a}{\delta^2} & -\dfrac{b}{\delta} \\ -\dfrac{b}{\delta} & b^2 \psi_1(a) \end{pmatrix}$$

and where ψ_1 is the tri-gamma function defined as the first derivative of the di-gamma function.

2.7.1.2. *Moments method estimator*

Here, we consider the estimator based on the moments method, only for the case of a homogeneous gamma process. The first increments are used to estimate t_0 and x_0, after estimating a and b using the other increments. When $t_0 \neq 0$ and $x_0 \neq 0$, we assume that $t_{i,1} = t_1$ for any $i \in \{1, \dots, n\}$, meaning that the first observation times is the same for all units (this particular situation has been also considered when looking at the inference for the Wiener process, see section 1.6.1).

Let $\mu_{(1)} = \mathbb{E}[X_{t_1}^{(i)}]$ and $\sigma_{(1)}^2 = \text{Var}[X_{t_1}^{(i)}]$, for any $i \in \{1, \dots, n\}$. We have:

$$\mu_{(1)} = \frac{a}{b}(t_1 - t_0) + x_0 \quad \text{and} \quad \sigma_{(1)}^2 = \frac{a}{b^2}(t_1 - t_0).$$

Let also

$$\mu_{(2)} = a/b = \mathbb{E}[Y_{i,j}]/s_{i,j} \quad \text{and} \quad \sigma_{(2)}^2 = a/b^2 = \text{Var}[Y_{i,j}]/s_{i,j},$$

where both quantities are independent of $(i,j) \in \{1, \dots, n\} \times \{1, \dots, m_i\}$.

Substituting $\mu_{(1)}$, $\sigma_{(1)}^2$, $\mu_{(2)}$ and $\sigma_{(2)}^2$ by their empirical estimators in the previous equations provides a set of four equations that are easily solved. We get:

$$\begin{cases} \widehat{t_0} = t_1 - \widehat{\sigma}_{(1)}^2/\widehat{\sigma}_{(2)}^2 \\ \widehat{x}_0 = \widehat{\mu}_{(1)} - \widehat{\mu}_{(2)}\widehat{\sigma}_{(1)}^2/\widehat{\sigma}_{(2)}^2 \\ \widehat{a} = \widehat{\mu}_{(2)}^2/\widehat{\sigma}_{(2)}^2 \\ \widehat{b} = \widehat{\mu}_{(2)}/\widehat{\sigma}_{(2)}^2 \end{cases}$$

where

$$\begin{cases} \widehat{\mu}_{(1)} = \dfrac{1}{n}\displaystyle\sum_{i=1}^{n} X_{i,1} \\ \widehat{\sigma}_{(1)}^2 = \dfrac{1}{n}\displaystyle\sum_{i=1}^{n}(X_{i,1} - \widehat{\mu}_{(1)})^2 \\ \widehat{\mu}_{(2)} = \dfrac{1}{n(\overline{m}-1)}\displaystyle\sum_{i=1}^{n}\sum_{j=2}^{m_i}\frac{Y_{i,j}}{s_{i,j}} \\ \widehat{\sigma}_{(2)}^2 = \dfrac{1}{n(\overline{m}-1)}\displaystyle\sum_{i=1}^{n}\sum_{j=2}^{m_i}\frac{(Y_{i,j} - s_{i,j}\widehat{\mu}_{(2)})^2}{s_{i,j}} \end{cases}$$

where $\overline{m} = \frac{1}{n}\sum_{i=1}^{n} m_i$ (see [BOR 15] where similar empirical quantities are introduced for a related model). This leads to asymptotically unbiased estimators that may however present some bias, when considering finite sample size. When the number n of individuals is "small", we could consequently prefer unbiased estimators of $\sigma^2_{(1)}$ and $\sigma^2_{(2)}$, which are the following one:

$$\begin{cases} \widetilde{\sigma}^2_{(1)} = \dfrac{1}{n-1}\displaystyle\sum_{i=1}^{n}(X_{i,1} - \widehat{\mu}_{(1)})^2 \\ \widetilde{\sigma}^2_{(2)} = \dfrac{1}{n(\bar{m}-1)-1}\displaystyle\sum_{i=1}^{n}\sum_{j=2}^{m_i}\dfrac{(Y_{i,j} - s_{i,j}\widehat{\mu}_{(2)})^2}{s_{i,j}} \end{cases}$$

Of course, $\widehat{\sigma}^2_{(i)}$ and $\widetilde{\sigma}^2_{(i)}$ are asymptotically equivalent for $i \in \{1,2\}$. Below, we go on with $\widehat{\sigma}^2_{(1)}$ and $\widehat{\sigma}^2_{(2)}$.

One can notice that $(\widehat{\mu}_{(1)}, \widehat{\sigma}^2_{(1)})$ and $(\widehat{\mu}_{(2)}, \widehat{\sigma}^2_{(2)})$ are two independent random vectors since they are defined on different independent increments. The following theorem states the asymptotic normality of $(\widehat{\mu}_{(1)}, \widehat{\sigma}^2_{(1)})$, which is given without proof since it is a classical result.

THEOREM 2.45.– Estimators $\widehat{\mu}_{(1)}$ and $\widehat{\sigma}^2_{(1)}$ are asymptotically normal:

$$\sqrt{n}\begin{pmatrix} \widehat{\mu}_{(1)} - \mu_{(1)} \\ \widehat{\sigma}^2_{(1)} - \sigma^2_{(1)} \end{pmatrix} \xrightarrow[n\to\infty]{d} \mathcal{N}(0, \Sigma_1)$$

where

$$\Sigma_1 = \begin{pmatrix} \dfrac{a}{b^2} & \dfrac{a(a+1)(a+2)}{b^3} \\ \dfrac{a(a+1)(a+2)}{b^3} & \dfrac{a((a+1)(a+2)(a+3)-a)}{b^4} \end{pmatrix}.$$

We now provide asymptotic results for $(\widehat{\mu}_{(2)}, \widehat{\sigma}^2_{(2)})$. Proofs are similar to that of Bordes *et al.* [BOR 15, theorems 2.1 and 2.3] devoted to a close degradation model and they are hence not given here. We begin with a strong consistency result.

THEOREM 2.46.– Under the following assumptions:

– (H_1) $\displaystyle\sum_{n\geq 1}\sum_{j=2}^{m_n} s_{n,j}^{-1}\left(\sum_{i=1}^{n} m_i - n\right)^{-2} < \infty$;

– (H_2) $\exists\delta_u, \forall i \in \mathbb{N}^\star; \forall j \in \{2, \ldots, m_i\}, s_{i,j} \leq \delta_u$, we have

$$\widehat{\mu}_{(2)} \xrightarrow[n\to\infty]{a.s.} \mu_{(2)} \quad \text{and} \quad \widehat{\sigma}^2_{(2)} \xrightarrow[n\to\infty]{a.s.} \sigma^2_{(2)}.$$

Next theorem states the asymptotic normality of $(\widehat{\mu}_{(2)}, \widehat{\sigma}^2_{(2)})$.

THEOREM 2.47.– If assumptions (H_1) and (H_2) hold and if the following assumption holds,

$$\text{– } (H_3)\ \forall u \in \{0,1\},\ \lim_{n\to\infty} \left(\sum_{i=1}^{n} m_i - n\right)^{-1} \sum_{i=1}^{n}\sum_{j=2}^{m_i} s_{i,j}^{u-2} = c_u < \infty,$$

then

$$\sqrt{n(\bar{m}-1)} \begin{pmatrix} \widehat{\mu}_{(2)} - \mu_{(2)} \\ \widehat{\sigma}^2_{(2)} - \sigma^2_{(2)} \end{pmatrix} \xrightarrow[n\to\infty]{d} \mathcal{N}(0, \Sigma_2)$$

where

$$\Sigma_2 = \begin{pmatrix} \dfrac{ac_1}{b^2} & \dfrac{2ac_1}{b^3} \\ \dfrac{2ac_1}{b^3} & \dfrac{6ac_1 + 3a^2}{b^4} \end{pmatrix}.$$

From theorems 2.45 and 2.47, we can notice that, in case of distinct m_i's, the normalization is different for the two sets of estimators. Therefore, it is not possible to derive an asymptotic behavior for the whole set of estimators in the general case. We now consider the specific case where all degradation processes are observed the same number m of times (but not necessarily at the same observation times). Hence, $m_i = m$ for any $i \in \{1, \ldots, n\}$ and $n(\bar{m}-1) = n(m-1)$. So, from theorem 2.47, we have:

$$\sqrt{n} \begin{pmatrix} \widehat{\mu}_{(2)} - \mu_{(2)} \\ \widehat{\sigma}^2_{(2)} - \sigma^2_{(2)} \end{pmatrix} \xrightarrow[n\to\infty]{d} \mathcal{N}\left(0, \frac{1}{m-1}\Sigma_2\right)$$

(where Σ_2 is provided in theorem 2.47). For such a sampling scheme, the δ-method (see theorem 3.1 in [VAA 98], for instance) can be used to prove the asymptotic normality of the estimator of (t_0, x_0, a, b), which provides the following result.

THEOREM 2.48.– Under assumptions (H_1)–(H_3), then

$$\sqrt{n}\begin{pmatrix} \widehat{t}_0 - t_0 \\ \widehat{x}_0 - x_0 \\ \widehat{a} - a \\ \widehat{b} - b \end{pmatrix} \xrightarrow[n\to\infty]{d} \mathcal{N}(0, G^T \Sigma G)$$

where

$$\Sigma = \begin{pmatrix} \Sigma_1 & 0 \\ 0 & \dfrac{1}{m-1}\Sigma_2 \end{pmatrix}$$

and

$$G = \begin{pmatrix} 0 & -\dfrac{1}{\sigma^2_{(2)}} & 0 & \dfrac{\sigma^2_{(1)}}{\sigma^4_{(2)}} \\ 1 & -\dfrac{\mu_{(2)}}{\sigma^2_{(2)}} & -\dfrac{\sigma^2_{(1)}}{\sigma^2_{(2)}} & \dfrac{\mu_{(2)}\sigma^2_{(1)}}{\sigma^4_{(2)}} \\ 0 & 0 & \dfrac{2\mu_{(2)}}{\sigma^2_{(2)}} & \dfrac{2\mu^2_{(2)}}{\sigma^4_{(2)}} \\ 0 & 0 & \dfrac{1}{\sigma^2_{(2)}} & \dfrac{2\mu_{(2)}}{\sigma^4_{(2)}} \end{pmatrix}$$

In the specific case where $t_0 = 0$ and $x_0 = 0$, the other two parameters are estimated as follows:

$$\widehat{a} = \widehat{\mu}^2/\widehat{\sigma}^2 \quad \text{and} \quad \widehat{b} = \widehat{\mu}/\widehat{\sigma}^2, \qquad [2.46]$$

where

$$\widehat{\mu} = \frac{1}{n\bar{m}} \sum_{i=1}^{n} \sum_{j=1}^{m_i} \frac{Y_{i,j}}{s_{i,j}}$$

and

$$\widehat{\sigma}^2 = \frac{1}{n\bar{m}} \sum_{i=1}^{n} \sum_{j=1}^{m_i} \frac{(Y_{i,j} - s_{i,j}\widehat{\mu})^2}{s_{i,j}}$$

with $\bar{m} = \frac{1}{n}\sum_{i=1}^{n} m_i$. Following the proof of the asymptotic normality of the MME for the general model, we can show the asymptotic normality of the estimators (with mostly identical assumptions as for the general case).

THEOREM 2.49.– Under the following assumptions:

– $(\tilde{\mathrm{H}}_1)$ $\displaystyle\sum_{n\geq 1}\sum_{j=1}^{m_n} s_{n,j}^{-1}\left(\sum_{i=1}^{n} m_i\right)^{-2} < \infty$;

– $(\tilde{\mathrm{H}}_2)$ $\exists d_u, \forall i \in \mathbb{N}^\star; \forall j \in \{1,\ldots,m_i\}, s_{i,j} \leq d_u$;

– $(\tilde{\mathrm{H}}_3)$ $\forall u \in \{0,1\}$, $\displaystyle\lim_{n\to\infty}\left(\sum_{i=1}^{n} m_i\right)^{-1}\sum_{i=1}^{n}\sum_{j=1}^{m_i} s_{i,j}^{u-2} = \tilde{c}_u < \infty$,

we have

$$\sqrt{n\bar{m}}\begin{pmatrix}\widehat{a}-a\\ \widehat{b}-b\end{pmatrix} \xrightarrow[n\to\infty]{d} \mathcal{N}(0, \tilde{G}^T\tilde{\Sigma}\tilde{G})$$

where

$$\tilde{\Sigma} = \begin{pmatrix}\dfrac{a\tilde{c}_1}{b^2} & \dfrac{2a\tilde{c}_1}{b^3}\\ \dfrac{2a\tilde{c}_1}{b^3} & \dfrac{6a\tilde{c}_1+3a^2}{b^4}\end{pmatrix}$$

and

$$\tilde{G} = \begin{pmatrix}\dfrac{2\mu}{\sigma^2} & \dfrac{2\mu^2}{\sigma^4}\\ \dfrac{1}{\sigma^2} & \dfrac{2\mu}{\sigma^4}\end{pmatrix} = \begin{pmatrix}2b & 2b^2\\ \dfrac{b^2}{a} & \dfrac{2b^3}{a}\end{pmatrix}.$$

To conclude, we discuss the assumptions involved in these theorems, which are not always that easy to interpret in practice. We here propose an alternate set of more intuitive (but stronger) assumptions introduced by Bordes *et al.* [BOR 15]:

– (A_1) same number of observations for all processes: for all $i \in \{1,\ldots,n\}$, $m_i = m$;

– (A_2) same instants of observations for all processes: for all $i \in \{1,\ldots,n\}$ and for all $j \in \{1,\ldots,m_i\}$, $t_{i,j} = t_j$;

– (A_3) regular instants (but not necessary identical instants for all processes): there exists δ such that, for any $i \in \{1,\ldots,n\}$ and for any $j \in \{1,\ldots,m_i\}$, $s_{i,j} = \delta$;

– (A_4) same time interval for all observations: there exists d such that, for all $i \in \{1,\ldots,n\}$, $t_{i,m_i} = d$;

– (A_5) uniformly bounded delay between consecutive observations: there exists $\delta_l > 0$ such that, for all $i \in \{1,\ldots,n\}$ and for all $j \in \{1,\ldots,m_i\}$, $s_{i,j} \geq \delta_l$.

One can notice that (A_2) implies (A_1). We can also easily check that (A_4) implies (H_2) and that jointly (A_1) and (A_5) imply (H_1). Moreover, simplifications may occur under some assumptions. For instance, if (A_3) and (A_4) hold, then (H_1) and (H_3) are equivalent to, respectively:

– (H_1') $\sum\limits_{n\geq 1} m_n^2 \left(\sum\limits_{i=1}^{n} m_i\right)^{-2} < \infty$;

– (H_3') $\forall u \in \{0,1\}$, $\lim\limits_{n\to\infty} \left(\sum\limits_{i=1}^{n} m_i\right)^{-1} \left(\sum\limits_{i=1}^{n} m_i^{3-u}\right) < \infty$.

In addition, if (A_1) holds, then (H_1') and (H_3') are satisfied. To finish, we now consider five different special cases that can be described in terms of assumptions (A_1)–(A_5):

1) same number of observations at the same regular instants over $[0,\delta]$: assumptions $(A_1) - (A_4)$ hold;

2) same number of observations at the same non-regular instants over $[0,\delta]$: assumptions (A_1), (A_2) and (A_4) hold;

3) $m_i = i$ and regular instants over $[0,\delta]$: assumptions (A_3) and (A_4) hold;

4) $m_i = i$ and regular instants over $[0,i\delta]$: assumptions (A_3) and (A_5) hold;

5) $m_i = 2^{i-1}$ and regular instants over $[0,\delta]$: assumptions (A_3) and (A_4) hold.

We can easily check that, for cases 1, 2 and 4, the estimators are consistent and asymptotically normal. Case 3 is sufficient to obtain consistency but it is not possible to derive asymptotic normality from the above theorems. Finally, under case 5, consistency and asymptotic normality of the estimators cannot be established based on the above limiting results.

REMARK 2.50.– The MME has the advantage of being explicit. It can be used as a starting point for computing the MLE from an iterative optimization procedure (Nelder–Mead algorithm, conjugate gradient method, etc.). Moreover, the asymptotic normality can be used to randomly draw initial starting points around the MME for a multi-start optimization procedure of the log-likelihood function.

2.7.2. *Other cases*

In most cases, only degradation data are available, as dealt with in the previous section. However, in some cases, we may also have failure data at our disposal, together or not with degradation data. These two cases are presented now, as well as a brief discussion on some mixtures of these situations.

2.7.2.1. *Estimation based only on failure data*

Consider now that only time-to-failure data are observed. Setting h to be the failure level, the distribution of the hitting time of the failure zone $[h, +\infty)$ has been previously derived in the case where $t_0 = 0$ and $x_0 = 0$ (see proposition 2.41 in section 2.6). For any $h > 0$, we set $\tilde{\tau}_h$ to be the hitting time of $[h, +\infty)$ by a standard gamma process $(\tilde{X}_t)_{t\geq 0}$ with shape function $A(\cdot;\theta)$ and rate parameter $b > 0$ (starting from $\tilde{X}_0 = 0$). For $x_0 < h$, let $\tau_{x_0,h}$ be the hitting of the same failure zone $[h, +\infty)$ by $(X_t)_{t\geq t_0}$ where, for any $t > t_0$, $X_t = x_0 + \tilde{X}_{t-t_0}$. Clearly, we have the following relationship between $\tau_{x_0,h}$ and $\tilde{\tau}_{h-x_0}$:

$$\tau_{x_0,h} = t_0 + \tilde{\tau}_{h-x_0}.$$

The probability density functions are hence linked by:

$$f_{\tau_{x_0,h}}(t) = f_{\tilde{\tau}_{h-x_0}}(t - t_0)\mathbf{1}_{t\geq t_0}.$$

Based on this equation, it is not possible to estimate both h and x_0 from failure data, but only $h - x_0$. Remembering that equation [2.25] provides the p.d.f. of $\tilde{\tau}_{h-x_0}$, it is easy to derive the p.d.f. of $\tau_{x_0,h}$. Assuming $\tau^{(1)}_{x_0,h}, \ldots, \tau^{(n)}_{x_0,h}$ to be a sample of $\tau_{x_0,h} = \tilde{\tau}_{h-x_0} + t_0$, we are hence able to estimate $(t_0, h - x_0, \theta, b)$ through the maximum likelihood method, where the log-likelihood function is given by:

$$L_2(t_0, h - x_0, \theta, b|\mathbf{D}_{obs}) = \sum_{i=1}^{n} \log f_{\tau_{x_0,h}}(\tau^{(i)}_{x_0,h}).$$

Note that there is no explicit expression for the MLE, which must be computed numerically.

2.7.2.2. *Estimation based on both degradation and failure data*

Now, we consider the case where, for the i-th component, we observe m_i degradation measurements and failure times: $(X^{(i)}_{t_{i,1}}, \ldots, X^{(i)}_{t_{i,m_i}}, \tau^{(i)}_{x_0,h})$. Given $\{X_{t_{i,m_i}} = x^{(i)}_{m_i}\}$ with $x^{(i)}_{m_i} < h$, then $\tau^{(i)}_{x_0,h}$ is conditionally identically distributed as $t_{i,m_i} + \check{\tau}^{(i)}_{h-x^{(i)}_{m_i}}$ where

$$\check{\tau}^{(i)}_{h-x^{(i)}_{m_i}} = \inf\left\{t \geq t_{i,m_i}; \check{X}^{(i)}_t \geq h - x^{(i)}_{m_i}\right\},$$

and $(\check{X}^{(i)}_t = (X_t - X_{t_{i,m_i}})\mathbf{1}_{t \geq t_{i,m_i}})_{t \geq 0}$ is a standard gamma process with shape parameter $\check{A}(t) = (A(t) - A(t_{i,m_i}))\mathbf{1}_{t \geq t_{i,m_i}}$ and rate parameter b. The p.d.f. of $\check{\tau}^{(i)}_{h-x^{(i)}_{m_i}}$ (and hence that of $\tau_{x^{(i)}_{m_i},h}$) is given by equation [2.24] or [2.25] with $\check{A}$ instead of A. The likelihood function is hence given by

$$L_3(t_0, x_0, \theta, b|\mathbf{D}_{obs}) = L_1(t_0, x_0, \theta, b|\mathbf{D}_{obs}) \prod_{i=1}^{n} f_{\tau_{x^{(i)}_{m_i},h}}(t^{(i)}_h - t_{i,m_i})$$

where L_1 is the first likelihood function (provided by equations [2.37], [2.38] or [2.41], according to the case). Here again, there is no explicit expression for the MLE.

2.7.2.3. *Mixture of various cases*

Of course, any situation can be considered, which combines all of the three above cases. Indeed, some components can be observed only through measurements, while failure times can be the only observations for others and/or both kinds of observations can be available for still another set of components. In such a situation, the likelihood function can be expressed in terms of products of functions L_1, L_2 and L_3.

2.8. Statistical inference in the semi-parametric model

We now aim at estimating $b \in \mathbb{R}$ and $A \in \mathcal{G}$, where $\mathcal{G}$ is the set of all real positive non-decreasing function A such that $A(0) = 0$ and $\lim_{t\to+\infty} A(t) = +\infty$ (functional parameter). Here, we only consider the non-parametric inference of a standard gamma process (i.e., with $x_0 = 0$ and $t_0 = 0$) based only on degradation data. Note that the semi-parametric inference is a more difficult task than the parametric inference and that we are only able to estimate A at inspection times. In the specific case where the inspection times are the same for all, say $t_1 < \cdots < t_m$, we can estimate $A(t_1), \ldots, A(t_m)$ or equivalently $\Delta A(t_1), \ldots, \Delta A(t_m)$ where $\Delta A(t_j) = A(t_j) - A(t_{j-1})$ for any $j \in \{1, \ldots, m\}$. More specifically, $\Delta A(t_1), \ldots, \Delta A(t_m)$ and b can be obtained by maximizing the likelihood function which turns to be in such a case:

$$L(\Delta A(t_1), \ldots, \Delta A(t_m), b|\mathbf{D}_{obs}) = \prod_{j=1}^{m}\prod_{i=1}^{n} \frac{b\,\Delta A(t_j)}{\Gamma(\Delta A(t_j))} Y_{i,j}^{\Delta A(t_j)-1} e^{-bY_{i,j}}.$$

The remainder of the section is devoted to the case where the inspection times are not the same within individuals and where the previous approach cannot be applied any more since consecutive intervals may overlap and so, alternative ways have to be developed. Such a problem has been studied by Wang [WAN 09a] and Ye *et al.* [YE 14]. The first paper uses the profile pseudo-likelihood function and the isotonic regression, while the second one is based on an EM algorithm. Note that Wang [WAN 09a] has also considered a model which includes covariates in the rate parameter of the gamma process (this point is not discussed here).

2.8.1. *Approach based on the pseudo-likelihood function*

Following [WAN 09a], we assume that X is observed as a random number M of times at instants $0 = T_0 < T_1 < \cdots < T_M$. Let $X_j = X(T_j)$ for all $j \in \{1, \ldots, M\}$, where $(M, T_1, \ldots, T_M)$ is assumed to be independent of X. Observations correspond to n i.i.d. copies

$$(M_i, T_{i,1}, \ldots, T_{i,M_i}, X_{i,1}, \ldots, X_{i,M_i})_{1\leqslant i\leqslant n}$$

of $(M, T_1, \ldots, T_M, X_1, \ldots, X_M)$. From these observations, we can consider successive increments (that are independent) and thus get the log-likelihood function defined by (up to an additive constant that does not depend on A and b):

$$\ell(A, b|\mathbf{D}_{obs}) = \sum_{i=1}^{n}\sum_{j=1}^{M_i} \{\Delta A_{i,j}(\log Y_{i,j} + \log b) - bY_{i,j} - \log\Gamma(\Delta A_{i,j})\}$$

where $Y_{i,j} = X_{i,j} - X_{i,j-1}$ and $\Delta A_{i,j} = A(T_{i,j}) - A(T_{i,j-1})$ for any $i \in \{1, \ldots, n\}$ and $j \in \{1, \ldots, M_i\}$. As explained in the introduction of this section in the general case, intervals $[T_{i,j-1}, T_{i,j})$ may overlap and so maximizing the log-likelihood function to estimate parameters is not possible. Hence, Wang [WAN 09a] has rather proposed to consider the pseudo-likelihood function obtained by ignoring the dependence between observations of the same degradation process. More precisely, the corresponding pseudo-log-likelihood function is given by

$$\tilde{\ell}(A, b|\mathbf{D}_{obs}) = \sum_{i=1}^{n}\sum_{j=1}^{M_i} \log f_{X_{i,j}}(X_{i,j}),$$

where we recall that, for all $i \in \{1, \ldots, n\}$ and for $j \in \{1, \ldots, M_i\}$, $f_{X_{i,j}}$ is the p.d.f. of $X_{i,j}$. Using the distribution of X_t for any $t \geq 0$, the pseudo-log-likelihood function writes (up to a non-informative additive constant):

$$\tilde{\ell}(A, b|\mathbf{D}_{obs}) = \sum_{i=1}^{n}\sum_{j=1}^{M_i} \{A(T_{i,j})(\log X_{i,j} + \log b) - bX_{i,j} - \log\Gamma(A(T_{i,j}))\}.$$

To maximize $\tilde{\ell}$, Wang has suggested to use a profile likelihood approach in two-step procedure. First it provides $\widetilde{A}_b$ that maximizes $\tilde{\ell}(\cdot, b|\mathbf{D}_{obs})$ as a function of b. The estimator $\widehat{b}$ of b is next obtained by maximizing $b \mapsto \tilde{\ell}(\widetilde{A}_b, b|\mathbf{D}_{obs})$.

As for the first step, let us introduce the q distinct observation time points $t_1 < \cdots < t_q$ and, for any given $b > 0$, let us see how to estimate A at these observation points. So, let $b > 0$ be fixed. The partial pseudo-log-likelihood $p\tilde{\ell}$ is given by

$$p\tilde{\ell}(A|\mathbf{D}_{obs}, b) = \sum_{i=1}^{n}\sum_{j=1}^{M_i} \{A(T_{i,j})(\log X_{i,j} + \log b) - \log\Gamma(A(T_{i,j}))\}.$$

It can be written as follows:

$$\tilde{p\ell}(A|\mathbf{D}_{obs}, b) = \sum_{l=1}^{q} w_l\{A(t_l)(u_l + \log b) - \log\Gamma(A(t_l))\}, \qquad [2.47]$$

where $w_1, \ldots, w_q$ counts the number of observations at each time point:

$$w_l = \sum_{i=1}^{n}\sum_{j=1}^{M_i} \mathbf{1}(T_{i,j} = t_l)$$

and where, for each $l \in \{1, \ldots, q\}$,

$$u_l = \frac{1}{w_l}\sum_{i=1}^{n}\sum_{j=1}^{M_i} \log(X_{i,j})\mathbf{1}(T_{i,j} = t_l).$$

Estimators $\widetilde{A}_b(t_1), \ldots, \widetilde{A}_b(t_q)$ can be obtained by maximizing the partial likelihood $\tilde{p\ell}$ under the constraint that $A(t_1) < \cdots < A(t_q)$ for $t_1 < \cdots < t_q$ (we recall that monotonicity is required for A). From the theory of monotone regression for exponential families (see [ROB 88], for instance), we have a characterization of the estimator $\widetilde{A}_b$ through the following lemma.

LEMMA 2.51.– Let $H^\star$ be the greatest convex minorant of points

$$\left\{\left(\sum_{j\le l} w_j, \sum_{j\le l} w_j u_j\right), 1 \le l \le q\right\}.$$

For each $l \in \{1, \ldots, q\}$, let z_l be the left derivative of $H^\star$ at $\sum_{j\le l} w_j$ and let

$$\widetilde{A}_b(t_l) = \psi_0^{-1}(z_l + \log b)$$

for each $b > 0$, where ψ_0^{-1} is the inverse of the di-gamma function. Then, for each b, $(\widetilde{A}_b(t_1), \ldots, \widetilde{A}_b(t_q))$ is the unique minimizer of [2.47] under the given monotonicity constraint.

Note that the greatest convex minorant H^* from the previous lemma can be obtained through the pool adjacent violators algorithm (PAVA) (see

[ROB 88], for instance). Now that $\widetilde{A}_b$ is constructed at the observation points t_i's for each $b > 0$, we recall that the estimator $\widehat{b}$ of b is obtained by maximizing $b \mapsto \tilde{\ell}(\widetilde{A}_b, b|\mathbf{D}_{obs})$. The last step consists of setting $\widehat{A}(t_l) = \widetilde{A}_{\widehat{b}}(t_l)$ for all $l \in \{1, \ldots, q\}$, which provides the final estimator at the observation time points. Between two successive observation points, $\widetilde{A}_b$ is linearly approximated, leading thus to a non-decreasing piecewise linear estimator with possible knots $t_1, \ldots, t_q$:

$$\forall l \in \{1, \ldots, q\}, \forall t \in [t_{l-1}, t_l),\ \widehat{A}(t) = \widehat{A}(t_{l-1}) + \frac{\widehat{A}(t_l) - \widehat{A}(t_{l-1})}{t_l - t_{l-1}}(t - t_{l-1})$$

with $t_0 = 0$ and, for all $t \geq t_q$, $\widehat{A}(t) = \widehat{A}(t_q)$. From the estimator $\widehat{A}$, we can derive an estimator of $a = A'$ on the time interval $[0, t_q)$. Quite naturally, $\widehat{a}$ is piecewise constant and is given by

$$\forall l \in \{1, \ldots, q\}, \forall t \in [t_{l-1}, t_l), \quad \widehat{a}(t) = \frac{\widehat{A}(t_l) - \widehat{A}(t_{l-1})}{t_l - t_{l-1}}. \qquad [2.48]$$

These estimators are consistent and the rate of convergence is of order $n^{-1/3}$. They are also asymptotically normal, but with a rate of convergence of order $n^{-1/2}$ (see theorems 1–3 in [WAN 09a]). However, in practice, it is simpler and more efficient to construct confidence intervals using bootstrap methods rather than using the asymptotic normality.

2.8.2. *Approach based on an EM algorithm*

Ye *et al.* [YE 14] have proposed an alternative way to estimate semi-parametrically the parameters of a gamma process, using an EM algorithm. In the balanced case, the problem is quite simple since the shape function can be estimated through its increments (between two successive observation times). We here go on with the unbalanced case, for which this is not feasible any more since increments may overlap. In that case, considering all the distinct observation times over all individuals, there are unobserved deterioration measures for some individuals at some of these times. These unobserved data can be viewed as missing data and can consequently be treated through an EM algorithm, which is known to be well suited and statistically efficient in such a situation (see, for instance, [MAC 08]). Here, the idea is to complete the data by reconstructing the missing ones.

Assume that we observe n items and that the degradation of the i-th item is measured m_i times at instants $t_{i,1} < \cdots < t_{i,m_i}$. We denote by $t_1 < \cdots < t_q$ the q distinct observation times over all the individuals. We also denote by $\mathbf{D}_{\text{obs}}$, $\mathbf{D}_{\text{full}}$ and $\mathbf{D}_{\text{miss}}$ the observed, complete and missing data, respectively:

$$\begin{aligned}\mathbf{D}_{\text{obs}} &= \bigcup_{i=1}^{n} \{(t_{i,j}, X_{t_{i,j}}^{(i)}), 1 \leq j \leq m_i\} \\ \mathbf{D}_{\text{full}} &= \bigcup_{i=1}^{n} \{(t_h, X_{t_h}^{(i)}), 1 \leq h \leq q\} \\ \mathbf{D}_{\text{miss}} &= \mathbf{D}_{\text{full}} \setminus \mathbf{D}_{\text{obs}}.\end{aligned}$$

The statistical problem is to estimate the set of parameters $\Xi = (A(t_1), \ldots, A(t_q), b) \in \mathbb{R}_+^{q+1}$. The EM algorithm is an iterative algorithm where the k-th iteration consists of the two following successive steps:

1) The E-step requires the computation of the Q-function defined as follows:

$$Q(\Xi|\Xi^{(k)}) = \mathbb{E}[\ell(\Xi; \mathbf{D}_{\text{full}})|\mathbf{D}_{\text{obs}}, \Xi^{(k)}]$$

where $\ell(\Xi; \mathbf{D}_{\text{full}})$ is the log-likelihood function based on the complete data $\mathbf{D}_{\text{full}}$.

2) The M-step aims at updating $\Xi^{(k+1)}$ by maximizing the Q-function:

$$\Xi^{(k+1)} = \text{argmax}_{\Xi \in \mathbb{R}^{q+1}} Q(\Xi|\Xi^{(k)})$$

We now give details on these two steps within an iteration:

– *E-step*: we start by writing the log-likelihood function when the complete data are available. For any $h \in \{1, \ldots, q\}$, we put $\Delta X_{i,t_h} = X_{t_h}^{(i)} - X_{t_{h-1}}^{(i)}$ (with $t_0 = 0$), which is gamma distributed with shape parameter $\Delta A_h = A(t_h) - A(t_{h-1})$ and scale parameter b: it corresponds to increments of degradation measures that could be observed or not. From the independence of the increments, it is straightforward that

$$\begin{aligned}\ell(\Xi; \mathbf{D}_{\text{full}}) &= nA(t_q) \log(b) - b \sum_{i=1}^{n} X_{t_q}^{(i)} \\ &\quad + \sum_{i=1}^{n} \sum_{h=1}^{q} [\Delta A_h \log \Delta X_{i,t_h} - \log \Gamma(\Delta A_h)],\end{aligned}$$

using that $A(t_q) = \Delta A_1 + \cdots + \Delta A_q$. From this expression of the log-likelihood, we have to compute the following expectations:

$$\forall i \in \{1, \ldots, n\}, \quad w_i^{(k)} = \mathbb{E}[X_{t_q}^{(i)} | \mathbf{D}_{obs}, \Xi^{(k)}]$$

and

$$\forall i \in \{1, \ldots, n\}, \forall h \in \{1, \ldots, q\}, \quad w_{i,h}^{(k)} = \mathbb{E}[\log \Delta X_{i,t_h} | \mathbf{D}_{obs}, \Xi^{(k)}].$$

For any $i \in \{1, \ldots, n\}$, we can decompose $X_{t_q}^{(i)}$ as the sum of two independent and gamma distributed random variables:

$$X_{t_q}^{(i)} = X_{t_{i,m_i}}^{(i)} + \left(X_{t_q}^{(i)} - X_{t_{i,m_i}}^{(i)} \right).$$

It follows that we have

$$\begin{aligned} w_i^{(k)} &= X_{t_{i,m_i}}^{(i)} + \mathbb{E}[X_{t_q}^{(i)} - X_{t_{i,m_i}}^{(i)} | \mathbf{D}_{obs}, \Xi^{(k)}] \\ &= X_{t_{i,m_i}}^{(i)} + \frac{1}{b^{(k)}} \left(A^{(k)}(t_q) - A^{(k)}(t_{i,m_i}) \right). \end{aligned}$$

Notice that, if $t_{i,m_i} = t_q$, then $w_i^{(k)} = X_{t_{i,m_i}}^{(i)}$. The expression of $w_{i,h}^{(k)}$ is given in the following theorem.

THEOREM 2.52.– [YE 14]

1) If the observations at times $t_{j+1}, \ldots, t_{j+h-1}$ are missing while the degradation measures at times t_j and t_{j+h} are available, then, for any $l \in \{1, \ldots, h\}$,

$$\begin{aligned} &\mathbb{E}[\log \Delta X_{i,t_{j+l}} | \mathbf{D}_{\text{obs}}, \Xi^{(k)}] = w_{i,j+l}^{(k)} \\ &= \log \left(X_{t_{j+h}}^{(i)} - X_{t_j}^{(i)} \right) + \psi_0 \left(\Delta A_{j+l}^{(k)} \right) - \psi_0 \left(A^{(k)}(t_{j+h}) - A^{(k)}(t_j) \right). \end{aligned}$$

2) If the observations at times $t_{j+1}, \ldots, t_q$ are missing while the degradation measure at time t_j is available, then, for any $l \in \{j, \ldots, q\}$,

$$\mathbb{E}[\log \Delta X_{i,t_l} | \mathbf{D}_{obs}, \Xi^{(k)}] = w_{i,l}^{(k)} = \psi_0(\Delta A_l^{(k)}) - \log b^{(k)}.$$

– *M-step*: now we have to maximize $Q(\Xi|\Xi^{(k)})$ with respect to Ξ. Starting from the expression of ℓ, we can derive the equations satisfied by $\Xi^{(k+1)}$. When differentiating ℓ with respect to b, we get

$$b^{(k+1)} = \frac{nA^{(k+1)}(t_q)}{\sum_{i=1}^{n} w_i^{(k)}}. \qquad [2.49]$$

For any $j \in \{1, \ldots, q\}$, when derivating with respect to ΔA_j, we have

$$\Delta A_j^{(k+1)} = \psi_0^{-1}\left(\frac{1}{n}\sum_{i=1}^{n} w_{i,j}^{(k)} - \log b^{(k+1)}\right).$$

We can recover the estimation of $A(t_q)$ at the k-th iteration by summing all the q above increments:

$$A^{(k+1)}(t_q) = \sum_{j=1}^{q} \Delta A_j^{(k+1)} = \sum_{j=1}^{q} \psi_0^{-1}\left(\frac{1}{n}\sum_{i=1}^{n} w_{i,j}^{(k)} - \log b^{(k+1)}\right). \qquad [2.50]$$

Hence, replacing in equation [2.50] the expression of $b^{(k+1)}$ obtained in equation [2.49], we can determine the equation satisfied by $A^{(k+1)}(t_q)$:

$$A^{(k+1)}(t_q) = \sum_{j=1}^{q} \psi_0^{-1}\left(\frac{1}{n}\sum_{i=1}^{n} w_{i,j}^{(k)} - \log n - \log A^{(k+1)}(t_q) + \log \sum_{i=1}^{n} w_i^{(k)}\right),$$

which is solved numerically. The estimator $\hat{b}^{(k+1)}$ is finally obtained through [2.49].

To conclude the presentation of the method, the whole procedure is summed up into the following algorithm.

ALGORITHM 2.53.– EM algorithm for the semi-parametric inference of a gamma process.

1) Parameter initialization: $\Xi^{(0)}$ is determined by assuming, for instance, that A is linear, say $A(t) = at$, using the results from section 2.7.1 and setting $A^{(0)}(t_j) = \widehat{a}\, t_j$ and $b^{(0)} = \widehat{b}$.

2) Repeat until convergence

a) $k \leftarrow k + 1$

b) *E-step*: computation of the Q-function:

$$Q(\Xi|\Xi^{(k)}) = nA(t_q^{(k)}) \log\left(b^{(k)}\right) - b^{(k)} \sum_{i=1}^{n} X_{t_{i,m_i}}^{(i)} + \sum_{i=1}^{n} \left(X_{t_q}^{(i)} - X_{t_{i,m_i}}^{(i)}\right)$$
$$+ \sum_{i=1}^{n} \sum_{h=1}^{q} \Delta A_h^{(k)} w_{i,h}^{(k)} - n \sum_{h=1}^{q} \log\Gamma(\Delta A_h^{(k)})$$

with $w_{i,h}^{(k)}$ given in theorem 2.52.

c) *M-step*: maximization of the Q-function:

– determine $A^{(k+1)}(t_q)$ as the solution of the following equation:

$$A^{(k+1)}(t_q) = \sum_{j=1}^{q} \psi_0^{-1}\left(\frac{1}{n}\sum_{i=1}^{n} w_{i,j}^{(k)} - \log n - \log A^{(k+1)}(t_q) + \log \sum_{i=1}^{n} w_i^{(k)}\right).$$

– deduce the value of $b^{(k+1)}$ that is equal to

$$b^{(k+1)} = \frac{nA^{(k+1)}(t_q)}{\sum_{i=1}^{n}\left[X_{t_{i,m_i}}^{(i)} + \left(A^{(k)}(t_q) - A^{(k)}(t_{i,m_i})\right)/b^{(k)}\right]}.$$

– compute the estimation of the increments of A:

$$\forall j \in \{1,\ldots,q-1\}, \quad \Delta A_j^{(k+1)} = \psi_0^{-1}\left(\frac{1}{n}\sum_{i=1}^{n} w_{i,j}^{(k)} - \log b^{(k+1)}\right).$$

– deduce the estimation of A at the q observation times:

$$\forall j \in \{1,\ldots,q\}, \quad A^{(k+1)}(t_j) = \sum_{h=1}^{j} \Delta A_h^{(k+1)}$$

3) Final estimator after $k^\star$ iterations: $\widehat{A}$ is estimated through a piecewise linear (non-decreasing) approximation:

$$\forall j \in \{1,\ldots,q\}, \forall t \in [t_{j-1}, t_j),$$
$$\widehat{A}(t) = A^{(k^\star)}(t_{j-1}) + \frac{A^{(k^\star)}(t_j) - A^{(k^\star)}(t_{j-1})}{t_j - t_{j-1}}(t - t_{j-1})$$

with $t_0 = 0$ and, for all $t \geq t_q$, $\widehat{A}(t) = A^{(k^\star)}(t_q)$.

An estimator of $a = A'$ can next be derived on the time interval $[0, t_q)$ as in [2.48].

2.9. Statistical inference for models with random effects (frailty model)

Sometimes, heterogeneity between individuals may occur. This can be translated through the introduction of a random parameter (or frailty variable) in the gamma process, where each individual corresponds to one realization of the random parameter. In this section, the rate parameter is assumed to be random: following Lawless and Crowder [LAW 04], it is further assumed to be gamma distributed. In [LAW 04], the authors provide estimators in the case where the shape function has a specified parametric form. Subsequently, Wang [WAN 08] has proposed a semi-parametric estimator based on a pseudo-likelihood function (in a similar manner as for the gamma process without random effects, see section 2.8.1). Next, Ye *et al.* [YE 14] have considered a semi-parametric estimator based on the EM algorithm. Below, we give some details on these estimators, first for the parametric model and next for the semi-parametric model.

Notice that, more recently, Pulcini [PUL 13] has adopted a different way to integrate random effects, assuming that the parameters involved in a given parametric shape function are random.

2.9.1. *Parametric model*

We here assume that $x_0 = 0$ and $t_0 = 0$ (standard gamma process). As proposed by Lawless and Crowder [LAW 04], let B be a gamma distributed random variable with shape parameter α and rate parameter β, whose p.d.f. is denoted by g. Given $B = b$, the process $(X_t)_{t \geq 0}$ is assumed to be conditionally distributed as a gamma process with shape function $A(\cdot)$ and rate parameter b. For any $t \geq 0$, we denote by $f_t(\cdot|b)$ the conditional probability density function of X_t given that $B = b$. The unconditional probability density function of X_t can easily be derived as follows:

$$f_t(x) = \int_0^\infty f_t(x|b)g(b)\mathrm{d}b \qquad [2.51]$$

$$= \int_0^\infty \frac{b^{A(t)}}{\Gamma(A(t))} x^{A(t)-1} e^{-bx} \frac{\beta^\alpha}{\Gamma(\alpha)} b^{\alpha-1} e^{-\beta b} \mathrm{d}b \qquad [2.52]$$

$$= \frac{\beta^\alpha x^{A(t)-1}}{\Gamma(A(t))\Gamma(\alpha)} \int_0^\infty b^{A(t)+\alpha-1} e^{-(x+\beta)b} \mathrm{d}b \qquad [2.53]$$

$$= \frac{1}{B(A(t),\alpha)} \frac{\beta^\alpha x^{A(t)-1}}{(x+\beta)^{A(t)+\alpha}}. \qquad [2.54]$$

The expectation of X_t is also straightforward to compute:

$$\mathbb{E}[X_t] = \mathbb{E}[\mathbb{E}[X_t|B]] = \mathbb{E}[A(t)B^{-1}] = A(t)\mathbb{E}[B^{-1}].$$

The random variable B^{-1} is said to be inverse gamma distributed. Its qth moment is finite as soon as $\alpha > q$, with:

$$\begin{aligned} \mathbb{E}[B^{-q}] &= \frac{\beta^\alpha}{\Gamma(\alpha)} \int_0^\infty y^{\alpha-q-1} e^{-\beta y} \mathrm{d}y \\ &= \beta^q \frac{\Gamma(\alpha-q)}{\Gamma(\alpha)} = \frac{\beta^q}{(\alpha-1)\cdots(\alpha-q)}. \end{aligned} \qquad [2.55]$$

In particular:

$$\mathbb{E}[X_t] = \frac{\beta}{\alpha-1} A(t), \quad \text{for } \alpha > 1. \qquad [2.56]$$

Consequently, the average degradation is still proportional to the shape function. Similarly, the second moment of X_t can be computed through:

$$\begin{aligned} \mathbb{E}[X_t^2] &= \mathbb{E}[\mathbb{E}[X_t^2|B]] = \{A(t) + A(t)^2\}\mathbb{E}[B^{-2}] \\ &= \{1 + A(t)\} \frac{\beta^2 A(t)}{(\alpha-1)(\alpha-2)}, \end{aligned}$$

provided that $\alpha > 2$. Under the same condition, its variance is thus equal to:

$$\mathrm{Var}(X_t) = \frac{\beta^2 A(t)}{(\alpha-1)^2(\alpha-2)} \{A(t) + \alpha - 1\}.$$

Hence, the variance of X_t is not proportional any more to the shape function. For instance, if we consider a linear function for A, then the expected degradation is linear while the variance of the degradation is quadratic.

For any $t \geq 0$, let $Z_t = \alpha X_t/(\beta A(t))$. The probability density function of Z_t is equal, for any $z \geq 0$, to

$$\begin{aligned} f_{Z_t}(z) &= \frac{\beta A(t)}{\alpha} f_t\left(\frac{\beta A(t) z}{\alpha}\right) \\ &= \frac{1}{B(A(t),\alpha)} \frac{1}{z} \left(\frac{A(t)z}{\alpha + A(t)z}\right)^{A(t)} \left(\frac{\alpha}{\alpha + A(t)z}\right)^{\alpha}, \end{aligned}$$

where we recognize a Fisher distribution with $(2A(t), 2\alpha)$ degrees of freedom. It can also be noticed that $X_t/(\beta+X_t)$ is beta distributed with parameters $A(t)$ and α (see in [JOH 95], p. 327, for instance). From this result, we can deduce the time-to-failure distribution for such a model. Indeed, using the same remark as for the previous model without random effects (see equation [2.27] in the proof of proposition 2.41), we have, for any $t \geq 0$,

$$\mathbb{P}[\tau_h > t] = \mathbb{P}[X_t < h] = \mathbb{P}[Z_t < \alpha h/(\beta A(t))] = F^{\text{Fish}}_{2A(t),2\alpha}(\alpha h/(\beta A(t))),$$

where $F^{\text{Fish}}_{2A(t),2\alpha}$ denotes the cumulative distribution function of the Fisher distribution with $(2A(t), 2\alpha)$ degrees of freedom (there is no explicit expression).

In order to obtain the likelihood function for the general sampling scheme, we need to compute the joint distribution of the increments (for one single individual). Of course, these increments are not independent since the random effect is common. However, given B, they are conditionally independent. Assume that X is observed m times at instants $t_1 < \cdots < t_m$. For any $j \in \{1, \ldots, m\}$, let $Y_j = X_{t_j} - X_{t_{j-1}}$ (with $t_0 = 0$ and $X_0 = 0$), $\Delta A_j = A(t_j) - A(t_{j-1})$ and $s_j = t_j - t_{j-1}$ (let us recall that $A(0) = 0$). Given B, the joint distribution of $(Y_1, \ldots, Y_m)$ is simply the product of the probability density functions. It follows that, for any $(y_1, \ldots, y_m) \in \mathbb{R}_+^m$, the joint probability density function $f_{Y_1,\ldots,Y_m}$ of $(Y_1, \ldots, Y_m)$ is

$$\begin{aligned} f_{Y_1,\ldots,Y_m}(y_1, \ldots, y_m) &= \int_0^{\infty} \prod_{j=1}^{m} f_{s_j}(y_j|b) g(b) \mathrm{d}b \\ &= \frac{\beta^{\alpha}}{\Gamma(\alpha)} \frac{\Gamma(A(t_m) + \alpha)}{(x_m + \beta)^{A(t_m)+\alpha}} \prod_{j=1}^{m} \frac{y_j^{\Delta A_j - 1}}{\Gamma(\Delta A_j)}, \end{aligned} \qquad [2.57]$$

similarly as for the computation of the p.d.f. of X_t (see equations [2.51]–[2.54]) and setting $x_m = y_1 + \cdots + y_m$. At last, we can remark that we recover the expression obtained before when $m = 1$ (observation of a single degradation). In addition, the covariance between two non-overlapping increments is also straightforward to compute by conditioning with respect to B and using the conditional independence of Y_j and Y_k as well as [2.56] and then [2.55]:

$$\begin{aligned} \text{Cov}(Y_j, Y_k) &= \mathbb{E}(\mathbb{E}(Y_j|B)\mathbb{E}(Y_k|B)) - \mathbb{E}(Y_j)\mathbb{E}(Y_k) \\ &= \Delta A_j \Delta A_k \left[\mathbb{E}(B^{-2}) - (\mathbb{E}(B^{-1}))^2\right] \\ &= \frac{\beta^2 \Delta A_j \Delta A_k}{(\alpha-1)^2(\alpha-2)}. \end{aligned}$$

From the joint distribution of the increments, we can provide the likelihood function for the general sampling scheme. Here, we assume a parametric model for A, which depends on some parameter $\theta \in \Theta \subseteq \mathbb{R}^p$ and, as previously, we denote by $A(\cdot;\theta)$ the shape function in this framework. Based on equation [2.57], we get the following expression:

$$L(\theta, \alpha, \beta|\mathbf{D}_{obs}) = \frac{\beta^{\alpha n}}{\Gamma(\alpha)^n} \prod_{i=1}^{n} \frac{\Gamma(A(t_{i,m_i};\theta)+\alpha)}{(x_{i,m_i}+\beta)^{A(t_{i,m_i};\theta)+\alpha}} \prod_{i=1}^{n}\prod_{j=1}^{m_i} \frac{y_{i,j}^{\Delta A_{i,j}(\theta)-1}}{\Gamma(\Delta A_{i,j}(\theta))}.$$

The MLE of (θ, α, β) can be computed numerically by maximizing the log likelihood:

$$\begin{aligned} \ell(\theta, \alpha, \beta|\mathbf{D}_{obs}) &= \alpha n \log\beta - n \log\Gamma(\alpha) \\ &+ \sum_{i=1}^{n} \left\{\log\Gamma(A(t_{i,m_i};\theta)+\alpha) - (A(t_{i,m_i};\theta)+\alpha)\log(x_{i,m_i}+\beta)\right\} \\ &+ \sum_{i=1}^{n}\sum_{j=1}^{m_i} \left\{\Delta A_{i,j}(\theta)\log y_{i,j} - \log\Gamma(\Delta A_{i,j}(\theta))\right\}. \end{aligned}$$

For some applications like the study of the remaining useful lifetimes for a given (observed) individual, it can be interesting to have an estimation of the random effects $b_1, \ldots, b_n$ for each unit. A possible estimator is

$$\widehat{b_i} = \frac{A(t_{i,m_i};\widehat{\theta})}{X_{i,m_i}}$$

where $\widehat{\theta}$ is the MLE obtained above. This is justified by the fact that $\mathbb{E}[X_{i,m_i}] = A(t_{i,m_i};\theta)/b_i$.

2.9.2. *Semi-parametric model*

As for the standard gamma process, semi-parametric estimation has been considered first by Wang [WAN 08] and next by Ye *et al.* [YE 14]: we aim at estimating $(\alpha, \beta) \in \mathbb{R}_+^2$ and $A \in \mathcal{G}$ where $\mathcal{G}$ is the set of all real positive non-decreasing functions A such that $A(0) = 0$ and $\lim_{t\to+\infty} A(t) = +\infty$ (functional parameter). The same approaches as for the standard gamma process have been used for this purpose (see section 2.8).

2.9.2.1. *Approach based on pseudo-likelihood*

We assume that X is observed as a random number M of times at instants $0 = T_0 < T_1 < \cdots < T_M$. Let $X_j = X_{T_j}$ for all $j \in \{1, \ldots, M\}$, where $(M, T_1, \ldots, T_M)$ is independent of X. Observations correspond to n i.i.d. copies

$$(M_i, T_{i,1}, \ldots, T_{i,M_i}, X_{i,1}, \ldots, X_{i,M_i})_{1 \leqslant i \leqslant n}$$

of $(M, T_1, \ldots, T_M, X_1, \ldots, X_M)$. The following semi-parametric estimator for this model has been studied by Wang [WAN 08].

As in section 2.8.1, we consider the pseudo-likelihood function, rather than the likelihood function. More precisely, up to a multiplicative constant (which does not depend on the unknown parameters), from equation [2.54], the pseudo-likelihood function is given by

$$\begin{aligned}\tilde{L}(A, \alpha, \beta | \mathbf{D}_{obs}) = & \prod_{i=1}^{n} \prod_{j=1}^{M_i} \frac{\Gamma(A(T_{i,j}) + \alpha)}{\Gamma(A(T_{i,j}))\Gamma(\alpha)} \\ & \times \left(\frac{X_{i,j}}{\beta + X_{i,j}} \right)^{A(T_{i,j})} \left(\frac{\beta}{\beta + X_{i,j}} \right)^{\alpha}.\end{aligned}$$

The pseudo-log-likelihood function is immediately derived:

$$\tilde{\ell}(A, \alpha, \beta | \mathbf{D}_{obs})$$
$$= \sum_{i=1}^{n} \sum_{j=1}^{M_i} \left\{ H_\alpha(A(T_{i,j})) - \log \Gamma(\alpha) + A(T_{i,j}) \log \frac{X_{i,j}}{\beta + X_{i,j}} \right.$$
$$\left. + \alpha \log \frac{\beta}{\beta + X_{i,j}} \right\},$$

where $H_\alpha(u) = \log \Gamma(u + \alpha) - \log \Gamma(u)$. We now derive the profile pseudo-log-likelihood function to estimate first $A(t_1), \ldots, A(t_q)$ where we denote by $t_1 < \cdots < t_q$ the set of the q distinct observation times. We have

$$p\tilde{\ell}(A | \mathbf{D}_{obs}, \alpha, \beta) = \sum_{l=1}^{q} w_l \{ H_\alpha(A(t_l)) + A(t_l) u_l \},$$

where, as before, $w_1, \ldots, w_q$ counts the number of observations at each time points:

$$w_l = \sum_{i=1}^{n} \sum_{j=1}^{M_i} \mathbf{1}(T_{i,j} = t_l)$$

and where, for any $l \in \{1, \ldots, q\}$,

$$u_l = \frac{1}{w_l} \sum_{i=1}^{n} \sum_{j=1}^{M_i} \log \frac{X_{i,j}}{\beta + X_{i,j}} \mathbf{1}(T_{i,j} = t_l).$$

The maximum of the above function can be obtained explicitly:

$$\widetilde{A}_{\alpha,\beta}(t_l)) = h_\alpha^{-1}(-z_l)$$

where

$$z_l = \min_{r \le l} \max_{s \ge l} \frac{w_r u_r + \cdots + w_s u_s}{w_r + \cdots + w_s}.$$

Here again, the computation of $z_1, \ldots, z_q$ can be obtained using the PAVA (see [ROB 88], for instance). Then, the estimator $(\widehat{\alpha}, \widehat{\beta})$ of (α, β) can be obtained by maximizing $(\alpha, \beta) \mapsto \widetilde{\ell}(\widetilde{A}_{\alpha,\beta}, \alpha, \beta | \mathbf{D}_{obs})$. Next, we set $\widehat{A}(t_l) = \widetilde{A}_{\widehat{\alpha},\widehat{\beta}}(t_l)$ for all $l \in \{1, \ldots, q\}$. The function A is finally constructed as a piecewise linear function.

Similarly as for the semi-parametric estimator based on the profile pseudo-likelihood for the standard gamma process, these estimators are consistent, still with a rate of convergence of order $n^{-1/3}$, and asymptotically normal, with a rate of convergence of order $n^{-1/2}$ (see theorems 1–3 in [WAN 08]). However, in practice, it is simpler and more efficient to construct confidence intervals using bootstrap methods rather than using the asymptotic normality.

2.9.2.2. *Approach based on the EM algorithm*

As for the standard gamma process (see section 2.8.2 for details), we can use the EM algorithm for a semi-parametric estimation of the parameters as suggested by Ye *et al.* [YE 14]. For such a model, the random effect is not observed and thus, even in the balanced case, it can be treated as a problem with missing data. More precisely, for the i-th unit, the missing data correspond to the random effect b_i and eventually the non-observed degradation measures, i.e. the measures at the instants $\{t_{i,j}, j \in \{1, \ldots, m_i\}\} \setminus \{t_1, \ldots, t_q\}$ (which can be an empty set, as in the balanced case, for instance). We denote by $\mathbf{D}_{\text{obs}}$, $\mathbf{D}_{\text{full}}$ and $\mathbf{D}_{\text{miss}}$, respectively, the observed data, the complete data and the missing data:

$$\begin{aligned}
\mathbf{D}_{\text{obs}} &= \bigcup_{i=1}^{n} \{(t_{i,j}, X_{t_{i,j}}^{(i)}), 1 \leq j \leq m_i\} \\
\mathbf{D}_{\text{full}} &= \bigcup_{i=1}^{n} \{(t_h, X_{t_h}^{(i)}), 1 \leq h \leq q\} \cup \{b_i\} \\
\mathbf{D}_{\text{miss}} &= \mathbf{D}_{\text{full}} \setminus \mathbf{D}_{\text{obs}}.
\end{aligned}$$

This statistical problem can be turned into estimating the following set of parameters $\Xi = (\alpha, \beta, A(t_1), \ldots, A(t_q)) \in \mathbb{R}_+^{q+2}$.

– *E-step*: this requires the computation of the Q-function defined as follows:

$$Q(\Xi | \Xi^{(k)}) = \mathbb{E}[\ell(\Xi; \mathbf{D}_{\text{full}}) | \mathbf{D}_{obs}, \Xi^{(k)}]$$

where $\ell(\Xi; \mathbf{D}_{\text{full}})$ is the log-likelihood function based on the complete data $\mathbf{D}_{\text{full}}$. The log-likelihood function (up to a constant) based on the complete data can be defined as the sum of two terms:

$$\ell(\Xi; \mathbf{D}_{\text{full}}) = \ell_1(A(t_1), \ldots, A(t_q); \mathbf{D}_{\text{full}}) + \ell_2(\alpha, \beta; \mathbf{D}_{\text{full}})$$

where

$$\begin{aligned}\ell_1(A(t_1), \ldots, A(t_q); \mathbf{D}_{\text{full}}) &= \sum_{i=1}^{n} \sum_{j=1}^{q} [\Delta A_j (\log \Delta X_{i,j} \\ &\quad + \log B_i) - \log \Gamma(\Delta A_j)]\end{aligned}$$

with $\Delta A_j = A(t_j) - A(t_{j-1})$ and $\Delta X_{i,j} = X_{t_j}^{(i)} - X_{t_{j-1}}^{(i)}$, and

$$\ell_2(\alpha, \beta; \mathbf{D}_{\text{full}}) = \sum_{i=1}^{n} [\alpha \log \beta + (\alpha - 1) \log B_i - \log \Gamma(\alpha) - \beta B_i] .$$

Hence, for the E-step of the k-th iteration, we need to compute the following three quantities defined for any $i \in \{1, \ldots, n\}$ and any $j \in \{1, \ldots, m_i\}$ as

$$\begin{aligned}\nu_i^{(k)} &= \mathbb{E}[B_i | \mathbf{D}_{\text{obs}}, \Xi^{(k)}] \\ \upsilon_i^{(k)} &= \mathbb{E}[\log B_i | \mathbf{D}_{\text{obs}}, \Xi^{(k)}] \\ \zeta_{i,j}^{(k)} &= \mathbb{E}[\log \Delta X_{i,j} | \mathbf{D}_{\text{obs}}, \Xi^{(k)}].\end{aligned}$$

First, for any $i \in \{1, \ldots, n\}$, we compute the conditional distribution of B_i given $\Delta X_{i,1}, \ldots, \Delta X_{i,m_i}$. For any $u \geq 0$, we have:

$$\begin{aligned}f_{B_i|\Delta X_{i,1},\ldots,\Delta X_{i,m_i}}(u|y_1, \ldots, y_{m_i}) &= \frac{f_{B_i,\Delta X_{i,1},\ldots,\Delta X_{i,m_i}}(u, y_1, \ldots, y_{m_i})}{f_{\Delta X_{i,1},\ldots,\Delta X_{i,m_i}}(y_1, \ldots, y_{m_i})} \\ &= \frac{f_{\Delta X_{i,1},\ldots,\Delta X_{i,m_i}|B_i}(y_1, \ldots, y_{m_i}|u) f_{B_i}(u)}{f_{\Delta X_{i,1},\ldots,\Delta X_{i,m_i}}(y_1, \ldots, y_{m_i})} \\ &= \frac{\left(X_{t_{i,m_i}}^{(i)} + \beta\right)^{A(t_{i,m_i})+\alpha}}{\Gamma(\alpha + A(t_{i,m_i}))} u^{A(t_{i,m_i})+\alpha-1} e^{-\left(X_{t_{i,m_i}}^{(i)} + \beta\right)u}\end{aligned} \quad [2.58]$$

since $X^{(i)}_{t_{i,m_i}} = \Delta X_{i,1} + \cdots + \Delta X_{i,m_i}$. Hence, conditionally to $\Delta X_{i,1}, \ldots, \Delta X_{i,m_i}$, B_i is gamma distributed with shape parameter $A(t_{i,m_i}) + \alpha$ and rate parameter $\beta + X^{(i)}_{t_{i,m_i}}$. We derive:

$$\nu_i^{(k)} = \frac{A^{(k)}(t_{i,m_i}) + \alpha^{(k)}}{\beta^{(k)} + X^{(i)}_{t_{i,m_i}}}.$$

From equation [4.352.1] in [GRA 07], we know that

$$\upsilon_i^{(k)} = \psi_0\left(A^{(k)}(t_{i,m_i}) + \alpha^{(k)}\right) - \log\left(\beta^{(k)} + X^{(i)}_{t_{i,m_i}}\right).$$

Note that Ye *et al.* [YE 14] have proposed an alternate computation based on the moment generating function of B_i and on its first derivative with respect to u at point $u = 0$.

Finally, to compute $\zeta_{i,j}^{(k)}$, we have to distinguish two cases for each unit: either the i-th unit is observed at times t_{j-1} and t_j, or not. In the former case, $\upsilon_{i,j}^{(k)} = \log Y_{i,j}$. In the latter case, if any of the observations at times t_{j-1} and t_j is missing, we have to compute $\zeta_{i,j}^{(k)}$ through a conditional expectation, which is given in the theorem below.

THEOREM 2.54.– [YE 14]

1) If the observations at times $t_{j+1}, \ldots, t_{j+h-1}$ are missing while the degradation measures at times t_j and t_{j+h} are available, then, for any $l \in \{1, \ldots, h\}$,

$$\begin{aligned}\mathbb{E}[\log \Delta X^{(i)}_{i,j+l} | \mathbf{D}_{\text{obs}}, \Xi] &= \log(X^{(i)}_{t_{j+h}} - X^{(i)}_{t_j}) \\ &\quad + \psi_0(\Delta A_{j+l}) - \psi_0(A(t_{j+h}) - A(t_j)).\end{aligned}$$

2) If the observations at times $t_{j+1}, \ldots, t_q$ are missing while the degradation measure at time t_j is available, then, for any $l \in \{j, \ldots, q\}$,

$$\mathbb{E}[\log \Delta X^{(i)}_{i,l} | \mathbf{D}_{\text{obs}}, \Xi] = \psi_0(\Delta A_l) - \psi_0(A(t_j) + \alpha) + \log(\beta + X^{(i)}_{t_j}).$$

– *M-step*: this aims at updating $\Xi^{(k+1)}$ by maximizing the Q-function as follows:

$$\Xi^{(k+1)} = \text{argmax}_{\Xi \in \mathbb{R}^{q+2}} Q(\Xi | \Xi^{(k)})$$

It can be noticed that ℓ_2 depends only on (α, β) and it corresponds to the likelihood function for a gamma distribution. It appears (see equations [2.45a] and [2.45b]) that $\alpha^{(k+1)}$ and $\beta^{(k+1)}$ are the solutions of the following equations:

$$\begin{cases} \log \beta^{(k+1)} - \psi_0(\beta e^{(k+1)}) = \log \bar{\nu}^{(k)} - \bar{\upsilon}^{(k)} \\ \alpha^{(k+1)} = \bar{\nu}^{(k)} \beta^{(k+1)} \end{cases}$$

where $\bar{\nu}^{(k)}$ and $\bar{\upsilon}^{(k)}$ are, respectively, the empirical means of $\nu_1^{(k)}, \ldots, \nu_n^{(k)}$ and of $\upsilon_1^{(k)}, \ldots, \upsilon_n^{(k)}$.

Next, we differentiate $\mathbb{E}[\ell_1(A(t_1), \ldots, A(t_q); \mathbf{D}_{\text{full}}) | \mathbf{D}_{\text{obs}}, \Xi]$ with respect to $A^{(k)}(t_j)$, for $j \in \{1, \ldots, q\}$ and it gives:

$$\Delta A^{(k+1)}(t_j) = \psi_0^{-1}\left(\bar{\upsilon}^{(k)} + \frac{1}{n}\sum_{i=1}^{n} \varsigma_{i,j}^{(k)}\right).$$

From this, we get:

$$A^{(k+1)}(t_j) = \sum_{h=1}^{j} \Delta A^{(k+1)}(t_h)$$

To conclude, a summary of the procedure is provided by the following algorithm.

ALGORITHM 2.55.– EM algorithm for the semi-parametric inference of a gamma process with random effect:

1) Parameter initialization: $\Xi^{(0)}$ is determined by assuming, for instance, that A is linear, say $A(t) = at$, using the results from section 2.7.1 and setting $A^{(0)}(t_j) = \widehat{a}\, t_j$ and $b^{(0)} = \widehat{b}$.

2) Repeat until convergence,

a) $k \leftarrow k+1$

b) Preliminary step. Computation of $\bar{\nu}^{(k)}$ and $\bar{\upsilon}^{(k)}$:

$$\bar{\nu}^{(k)} = \frac{1}{n}\sum_{i=1}^{n} \frac{A^{(k)}(t_{i,m_i}) + \alpha^{(k)}}{\beta^{(k)} + X_{t_{i,m_i}}^{(i)}}$$

and

$$\bar{\upsilon}^{(k)} = \frac{1}{n}\sum_{i=1}^{n} \left[\psi_0\left(A^{(k)}(t_{i,m_i}) + \alpha^{(k)}\right) - \log\left(\beta^{(k)} + X_{t_{i,m_i}}^{(i)}\right)\right].$$

c) E-step. Computation of the Q-function:

$$Q(\Xi|\Xi^{(k)}) = \mathbb{E}[\ell_1(A(t_1), \dots, A(t_q); \mathbf{D}_{\text{full}})|\mathbf{D}_{\text{obs}}, \Xi^{(k)}] + \mathbb{E}[\ell_2(\alpha, \beta; \mathbf{D}_{\text{full}})|\mathbf{D}_{\text{obs}}, \Xi^{(k)}]$$

where

$$\mathbb{E}[\ell_1(A(t_1), \dots, A(t_q); \mathbf{D}_{\text{full}})|\mathbf{D}_{\text{obs}}, \Xi^{(k)}] = \sum_{i=1}^{n}\sum_{j=1}^{q} \Delta A_j^{(k)} \zeta_{i,j}^{(k)} + nq\bar{\nu}^{(k)} - n\sum_{j=1}^{q} \log \Delta A_j^{(k)}$$

with $\zeta_{i,j}^{(k)}$ given in theorem 2.54 and where

$$\mathbb{E}[\ell_2(\alpha, \beta; \mathbf{D}_{\text{full}})|\mathbf{D}_{\text{obs}}, \Xi^{(k)}] = n\alpha^{(k)} \log \beta^{(k)} - n \log \Gamma(\alpha^{(k)}) + n(\alpha^{(k)} - 1)\bar{\upsilon}^{(k)} - n\beta^{(k)}\bar{\nu}^{(k)}.$$

d) M-step. Maximization of the Q-function:

– determine $\beta^{(k+1)}$ as the solution of the following equation:

$$\log \beta^{(k+1)} - \psi_0(\beta^{(k+1)}) = \log \bar{\nu}^{(k)} - \bar{\upsilon}^{(k)}$$

– deduce the value of $\alpha^{(k+1)}$ that is equal to

$$\alpha^{(k+1)} = \bar{\nu}^{(k)} \beta^{(k+1)}$$

– compute the estimation of the increments of A:

$$\forall j \in \{1, \ldots, q\}, \quad \Delta A^{(k+1)}(t_j) = \psi_0^{-1}\left(\bar{\upsilon}^{(k)} + \frac{1}{n}\sum_{i=1}^{n} \varsigma_{i,j}^{(k)}\right).$$

– deduce the estimation of A at the q observation times:

$$\forall j \in \{1, \ldots, q\}, \quad A^{(k+1)}(t_j) = \sum_{h=1}^{j} \Delta A^{(k+1)}(t_h)$$

3) Final estimator after $k^\star$ iterations: once $A^{(k^\star)}(t_1), \ldots, A^{(k^\star)}(t_q)$ are computed, the final estimator $\widehat{A}(t)$ is constructed as a piecewise linear estimator on $[0, t_q)$, as in algorithm 2.53.

An estimator of $a = A'$ can also be derived as in [2.48].

2.10. Extensions and related models

In conclusion, we here mention a few other models from the literature, whose constructions are based on gamma processes. Note that we do not intend to give an exhaustive state of art on the subject, but only want to draw the reader's attention to some other interesting models from the literature, which are not discussed in this book.

As a first step, we describe two other ways of introducing covariates or accelerating factors into a gamma process. Next, we briefly present the case of a gamma process perturbed by an independent Brownian motion. This model may be of particular interest for selection model purpose. At last, we introduce several extensions of the gamma process, which are more flexible for modeling purpose, with the price of a higher technicality, however.

2.10.1. *Accelerated gamma process*

In section 2.9, we have considered a gamma process with a random rate parameter. Introducing random effects (or frailty parameter) in a degradation model allows to take into account a unit to unit variability. The source of this variability could be induced by the environmental conditions of observation, which may be different from units to units. Alternatively, we can consider a degradation model where the parameters may depend on some covariates that can also reflect the environment. These covariates are generally assumed to be deterministic and known for each observed unit. They can be seen as a way to include an accelerating factor within a degradation model. The idea is that collecting degradation measurements from units observed under a higher stress can be done more quickly than under a nominal condition. When the units are observed in an accelerated but controlled environment, this can be used for extrapolating the degradation behavior of units under the nominal condition.

In [PAR 05], Park and Padgett consider a (standard) gamma process whose shape parameter is given by $A(t) = a_L t$ for any $t \geq 0$, where a_L depends on an acceleration variable (or stress level) L. Several parametric link functions a_L are reviewed in [PAR 05]:

- power law: $a_L = \xi L^\eta$;
- Arrhenius law: $a_L = \xi e^{\eta/L}$;
- inverse-log law: $a_L = \xi(\log L)^\eta$;
- exponential law: $a_L = \xi e^{\eta L}$;
- inverse–linear law: $a_L = \xi + \eta L$.

Based on successive observations of units at distinct constant acceleration levels, Park and Padgett [PAR 05] provides the likelihood function, which is next numerically optimized to get the MLE. The time-to-failure distribution is also computed. The approach is illustrated on carbon film resistors data.

More recently, Ling *et al.* [LIN 14] have considered a non-increasing deterioration model based on a gamma process, which is obtained through a logarithmic transformation of the data followed by a time scaling, as proposed by Whitmore and Schenkelberg [WHI 97]. An acceleration model is proposed where, under a constant stress level L, the random variable $\log(X_t) - \log(X_{t+s})$ is gamma distributed $\mathcal{G}(e^a L^b s, e^u L^v)$ (with $0 < s, t$). The link function between the parameters of the gamma distribution and the stress level L is known as the inverse power law, which is commonly used for electrical current data, for instance [ESC 06]. Contrary to the previously described model from [PAR 05], the accelerating factor L acts here both on the scale and shape functions. For this model, the authors have written the likelihood function (based on a similar sampling scheme as [PAR 05]) and studied several quantities of interest such as the mean and median time to failure, the remaining useful life, etc.

2.10.2. *Perturbed gamma process*

Bordes *et al.* [BOR 15] have considered a gamma process perturbed by an independent Brownian motion as follows:

$$\forall t \geq 0, \quad X_t = Z_t + \sigma B_t,$$

where $\sigma \in \mathbb{R}$, $(Z_t)_{t\geq 0}$ is a standard gamma process and $(B_t)_{t\geq 0}$ is a Brownian motion independent of $(Z_t)_{t\geq 0}$. For such a model, they propose a parametric estimator based on the moments method. This estimator is proved to be consistent and asymptotically normal, under a general sampling scheme. A first interest of this model is that it allows us to fit a data set that may contain negative increments. From a practical point of view, such a situation may be due to measurements errors or to unreported maintenance actions for instance. A second interest is that the perturbed gamma process contains both the Wiener process (when $\sigma = 0$) and the gamma process (when a/b tends to $\mu > 0$ and a/b^2 tends to zero) as special cases. This allows us to test the Wiener or the gamma process against the perturbed gamma process, and hence helps to select the best adapted model. The time-to-failure distribution for this model has been studied by Paroissin and Rabehasaina [PAR 15].

2.10.3. *Extensions of the gamma process*

A notable limitation of a standard Gamma process is that its variance-to-mean ratio is constant over time, which may be restrictive within an applicative context, see [GUI 12] for a practical example. To overcome this restriction, it is possible to use an extended version of a gamma process (extended gamma process: EGP) where the rate parameter is allowed to vary over time. These processes have been introduced mostly simultaneously by [ÇIN 80] and [DYK 81]. Note that EGPs are also called weighted gamma processes in the literature [ISH 04]. An EGP can be constructed as a stochastic integral with respect to a standard gamma process: starting from a standard gamma process $(X_t)_{t\geq 0}$ with $A(.)$ as shape function and 1 as (constant) rate parameter, an EGP $(Y_t)_{t\geq 0}$ with the same shape function $A(.)$ and some positive function $b(.)$ as *rate function* is constructed through the following:

$$Y_t = \int_{(0,t]} \frac{dX_s}{b(s)} = \sum_{s\leq t} \frac{\Delta X_s}{b(s)}, \quad \text{for a.s. all } t > 0, \qquad [2.59]$$

where the last equality relies on the fact that the number of jumps of the standard gamma process $(X_t)_{t\geq 0}$ is almost surely countable (see equation [2.15]) and where $\Delta X_s = X_s - X_{s-}$ stands for the size of the jump at time s. An EGP still has independent increments and it is infinitely divisible [ÇIN 80]. Also, an explicit formula is available for its Laplace transform, with

$$\mathcal{L}_{Y_t}\lambda := \mathbb{E}\left(e^{-\lambda(Y_t)}\right) = \exp\left(-\int_{(0,t]} \log\left(1 + \frac{\lambda}{b(s)}\right) da(s)\right),$$

for all $t, \lambda \geq 0$. Its mean and variance are given by

$$\mathbb{E}(Y_t) = \int_{(0,t]} \frac{da(s)}{b(s)} \text{ and } \mathbb{V}(Y_t) = \int_{(0,t]} \frac{da(s)}{b(s)^2}.$$

The distribution of Y_t is not available in full form but tools are developed in [ALM 15] for its numerical computation (and approximate simulation). A discrete-time version of an EGP is proposed and used for deterioration modeling in [GUI 12]. In [LAU 96, ISH 04], EGPs are used for Bayesian modeling of the hazard function.

Just as for a classical gamma process, the conditional distribution of an EGP increment $Y_{t+h} - Y_t$ given all the past of the process up to time t still depends only on the time t and not on the deterioration level Y_t. This can be a restriction within an applicative context. To overcome this, a further extension of the gamma process named transformed gamma process has been proposed in [GIO 15], "where the distribution of the degradation increment in a future time interval depends both on the current age and the current degradation level". We refer to [GIO 15] for details.

3

Doubly Stochastic Marked Poisson Processes

In this chapter, another type of degradation process $(X_t)_{t\geq 0}$ is considered, whose paths are monotone non-decreasing step functions. The starting point is a doubly stochastic Poisson process (T_n), whose intensity function is random. Each arrival time T_n induces an instantaneous non-negative jump of size Z_n for the degradation level, which is constant between jumps. Then, X_t stands for the total degradation accumulated over jumps up to time t. The size Z_n of a jump may depend on the jump time T_n, leading to a so-called position-dependent marking of the doubly stochastic Poisson process (T_n). The process $\mathbf{N} = ((T_n, Z_n))_{n\geq 1}$ can be seen as a marked point process, which is presented in detail, e.g. in [LAS 95] or [AND 93]. The doubly stochastic Poisson process was introduced by Cox [COX 55]. Cramer [CRA 69] applied it in risk theory, and Grandell [GRA 91] gave a detailed discussion of this process and of its impact in risk theory. Further applications of the doubly stochastic Poisson process (T_n) may be found in reliability theory, medicine and queueing theory [BRE 81, AND 93, GRA 97, KON 92]. Note that contrary to the previous chapters, the cumulative process (X_t) is a Lévy process only in very specific cases.

As already mentioned, each arrival time T_n entails a sudden increase in the deterioration level, leading to a so-called cumulative shock model. There is a stream of papers concerned with shock models. Applications were given – among others – by Sobczyk [SOB 87] and Esary *et al.* [ESA 73]. Further papers dealing with shock models are in [FEN 94, SHA 83, AVE 98] and [WEN 99]. For the doubly stochastic Poisson process, we assume that the cumulative intensity of (T_n) is a random multiple Y of a deterministic function η. Frequently, these stochastic intensities are more realistic than the deterministic intensity of a Poisson process. Each realization of the random variable Y leads to a different individual intensity for (T_n), which could be the result of distinct environmental conditions for the different items or of some personal frailty for each of them.

Different parametric models are envisioned in the following, both for the deterministic part η and random multiple Y of the intensity of the doubly stochastic Poisson process as well as for the distribution of the marks. Our aim is to describe examples of suitable models for degradation accumulation and to calculate the distribution of the corresponding cumulative degradation. We also consider the first passage time and determine its distribution.

Furthermore, estimators for the parameters of the deterministic part, the distribution of Y, and the distribution of the marks Z_n $(n \geq 1)$ are determined under different levels of observable information. We consider maximum likelihood estimators as well as moment estimators and compare these two methods.

3.1. Modeling degradation by marked point processes

Let $(\Omega, \mathcal{F}, \mathbb{P})$ be a fixed probability space and let $(\mathcal{F}_t)$ be a filtration in $\mathcal{F}$. The random variable T_n $(n \geq 1)$ is the time of the n-th shock. We suppose

$$T_n < T_{n+1} \text{ a.s. if } T_n < \infty \text{ and } T_n = T_{n+1} = \infty \quad \text{otherwise.}$$

The size of the n-th increment of the cumulative degradation process $(X_t)_{t \geq 0}$ is given by a non-negative random variable Z_n $(n \geq 1)$. Thus,

$$X_t = \sum_{n=1}^{\infty} Z_n \cdot \mathbf{1}_{\{T_n \leq t\}} = \sum_{n=1}^{N_t} Z_n$$

describes the total amount of degradation at time t. The sequence $\mathbf{N} = ((T_n, Z_n))$ is called a *marked point process*, and $N_t = \max\{i \geq 1 : T_i \leq t\}$ is the random variable representing the number of shocks occurred up to time t. If the distribution of the marks Z_n is independent of (T_n), then the conditional c.d.f. of the cumulative degradation at time t given N_t is provided by the convolution $F * (\mathrm{d}F)^{*(N_t - 1)}$, where F is the common c.d.f. of all Z_n's and $(\mathrm{d}F)^{*(n-1)}$ represents the $(n-1)$-th convolution of the corresponding probability measure $\mathrm{d}F$. Frequently, it is of interest to discuss the first passage problem that the process (X_t) exceeds a pre-specified constant threshold level $h > 0$ for the first time. In the most general case, the system can start from a (random or deterministic) initial state X_0 at time $T_0 := 0$. Then, the corresponding first passage time τ_h is given as

$$\tau_h = \inf\{t : X_0 + \sum_{n=1}^{\infty} Z_n \mathbf{1}_{\{T_n \leq t\}} \geq h\}. \qquad [3.1]$$

Let us mention that τ_h coincides with some T_m for $m \in \mathbb{N}$. Typical courses of realizations of the processes N_t and Z_t as well as of the variable τ_h are sketched in Figure 3.1 for $X_0 = 0$.

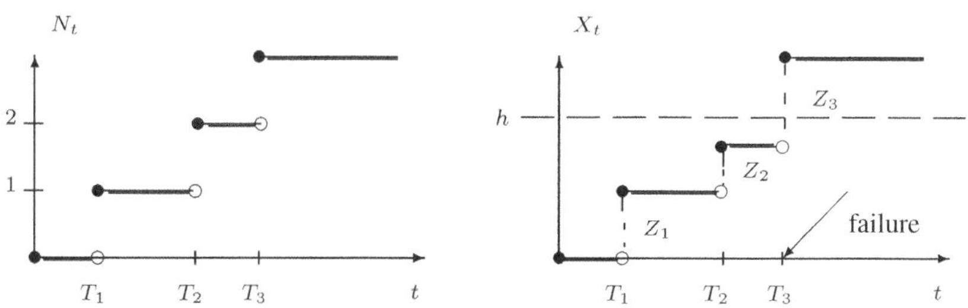

Figure 3.1. *A realization of N_t and X_t*

3.1.1. *The sequence* $(T_n)_{n\geq 1}$

First, we consider parametric models for the sequence of shocks $(T_n)_{n\geq 1}$. Let be given the filtration $(\mathcal{F}_t) = (\mathcal{F}_t^{\mathbf{N}}) \vee \sigma(Y)$, where $(\mathcal{F}_t^{\mathbf{N}})$ describes the internal history of $\mathbf{N}$, and Y (generating $\mathcal{F}_0$) is a non-negative random variable with finite expectation. The cumulative $(\mathbb{P}, \mathcal{F}_t)$-stochastic intensity $\bar{\nu}(t)$ of (T_n) is assumed to be given by $\bar{\nu}(t) = Y \cdot \eta(t)$, where $\eta(t)$ is a deterministic function with derivative $\xi(t) \geq 0$. Hence, given the outcome $Y = y$, the random variable N_t is Poisson distributed with mean $y \cdot \eta(t)$. And the unconditional distribution of N_t is given by

$$p_k(t) = \mathbb{P}(N_t = k) = \mathbb{E}\left(\frac{(Y\eta(t))^k}{k!} \exp(-Y\eta(t))\right), \quad k = 0, 1, \ldots. \qquad [3.2]$$

Based on this equation, the distribution of N_t appears as a mixture of Poisson distributions (where the distribution of Y is the mixing distribution) and the sequence $(T_n)_{n\geq 0}$ is sometimes called a *mixed Poisson process* or *Cox process*. Another vocabulary is used in the sequel, which is provided by the following definition.

DEFINITION 3.1.– *The sequence* $(T_n)_{n\geq 1}$ *with* $T_n = \min\{t \geq 0 : N_t = n\}$ *is called a* $(\mathbb{P}, \mathcal{F}_t)$*-doubly stochastic Poisson process.*

Both the deterministic function η and the distribution of Y can be specified: the most common models for η are

$$\eta(t) = t^{\alpha+1},\ \alpha > -1 \qquad \text{(Weibull type)},$$

$$\eta(t) = t \cdot e^{\alpha t^{\gamma}},\ \alpha \geq 0, \gamma \geq 0 \qquad \text{(log-linear type)},$$

$$\eta(t) = t \cdot (1 + \log(1 + \alpha t^{\gamma})),\ \alpha \geq 0, \gamma > -1 \qquad \text{(logistic type)}.$$

Taking $\alpha = 0$ in the previous deterministic functions provides the specific case where $\eta(t) = t$ and where given the outcome $Y = y$, the process $(N_t)_{t\geq 0}$ is a homogeneous Poisson process with parameter y.

The random variable Y is a frailty variable. The notion of frailty provides a convenient way to introduce random effects and unobserved heterogeneity into models for survival and degradation data. In its simplest form, a frailty is an unobserved random proportionality factor that modifies the hazard function of an individual or of related individuals. In the simplest case where $\mathbb{P}(Y = y_0) = 1$, the sequence $(T_n)_{n\geq 0}$ represents a *non-homogeneous Poisson process* with the deterministic intensity function $y_0 \cdot \xi(t)$ where $\xi(t) = \eta'(t)$. Three more sophisticated examples are now considered.

EXAMPLE 3.2.– Let Y be uniformly distributed in $[a, b]$ with $0 \leq a < b$. Equation [3.2] first provides

$$p_k(t) = \int_a^b \left(\frac{(y\eta(t))^k}{k!} \exp(-y\eta(t)) \right) \frac{1}{b-a} \mathrm{d}y. \qquad [3.3]$$

Assuming $\eta(t) > 0$, setting $z = \eta(t)y$ and using an integration by parts, we now have:

$$\int_a^b \frac{y^k}{k!} e^{-\eta(t)y} \mathrm{d}y$$

$$= \frac{1}{\eta(t)^{k+1}} \int_{\eta(t)a}^{\eta(t)b} \frac{z^k}{k!} e^{-z} \mathrm{d}z$$

$$= \frac{1}{\eta(t)^{k+1}} \left(\frac{(\eta(t)a)^k}{k!} e^{-\eta(t)a} - \frac{(\eta(t)b)^k}{k!} e^{-\eta(t)b} \right.$$

$$\left. + \int_a^b \frac{z^{k-1}}{(k-1)!} e^{-z} \mathrm{d}z \right)$$

$$= \frac{1}{\eta(t)^{k+1}} \sum_{i=0}^{k} \left(\frac{(\eta(t)a)^i}{i!} e^{-\eta(t)a} - \frac{(\eta(t)b)^i}{i!} e^{-\eta(t)b} \right)$$

based on induction for the last line. Starting again from [3.3], we derive

$$p_k(t) = \frac{1}{\eta(t)\,(b-a)} \sum_{i=0}^{k} \left(\frac{(a\,\eta(t))^i}{i!} e^{-a\,\eta(t)} - \frac{(b\,\eta(t))^i}{i!} e^{-b\,\eta(t)} \right). \quad [3.4]$$

EXAMPLE 3.3.– Let $Y - y_0$ be $\mathcal{G}(b, c)$-distributed with $c > 0$ and $b > 0$. Based on [3.2], we have to calculate

$$p_k(t) = \int_{y_0}^{\infty} \frac{(y\eta(t))^k}{k!} e^{-y\eta(t)} \frac{c^b}{\Gamma(b)} (y - y_0)^{b-1} e^{-c(y-y_0)} \mathrm{d}y.$$

Setting $z = y - y_0$ and using the binomial theorem for the second line, we get

$$p_k(t) = \frac{c^b(\eta(t))^k}{\Gamma(b)\,k!} e^{-y_0\eta(t)} \int_0^{\infty} (z + y_0)^k e^{-z\eta(t)} z^{b-1} e^{-cz} \mathrm{d}z$$

$$= \frac{c^b(\eta(t))^k}{\Gamma(b)\,k!} e^{-y_0\eta(t)} \int_0^{\infty} \sum_{i=0}^{k} \binom{k}{i} z^i y_0^{k-i} e^{-z\eta(t)} z^{b-1} e^{-cz} \mathrm{d}z$$

$$= \frac{c^b(\eta(t))^k}{\Gamma(b)\,k!} e^{-y_0\eta(t)} \sum_{i=0}^{k} \binom{k}{i} y_0^{k-i} \frac{\Gamma(b+i)}{(c+\eta(t))^{b+i}} \cdot I(i),$$

where, for each $0 \leq i \leq k$, $I(i)$ is an integral over the p.d.f. of a $\mathcal{G}(b + k - i, c + \eta(t))$ distribution:

$$I(i) = \int_0^{\infty} \frac{(c + \eta(t))^{b+i}}{\Gamma(b+i)} z^{b+i-1} e^{-(c+\eta(t))\cdot z} \mathrm{d}z = 1.$$

Rearranging the terms provides

$$p_k(t) = \sum_{i=0}^{k} \frac{\Gamma(b+i)}{\Gamma(b)i!(k-i)!}(y_0\eta(t))^{k-i}\left(\frac{c}{c+\eta(t)}\right)^b\left(\frac{\eta(t)}{c+\eta(t)}\right)^i e^{-y_0\eta(t)}, \tag{3.5}$$

where we recognize the Delaporte distribution (see, e.g. [VOS 08], pp. 618–619), which is a discrete probability distribution with applications in actuarial science. The Delaporte distribution is a special case of a so-called *mixed Poisson distribution*, i.e. a Poisson distribution where the mean parameter is random. For the Delaporte distribution, we assume that the mean parameter is a sum of a fixed component λ, and a $\mathcal{G}(\alpha, 1/\beta)$ distributed random component. The probabilities of the Delaporte distribution are then given by

$$p_k = \sum_{i=0}^{k} \frac{\Gamma(\alpha+i)\beta^i\lambda^{k-i}e^{-\lambda}}{\Gamma(\alpha)i!(1+\beta)^{\alpha+i}(k-i)!}, i = 0, 1, \ldots$$

In our case, we have a fixed component $y_0\eta(t)$ and a $\mathcal{G}(b, \frac{c}{\eta(t)})$ random component and [3.5] appears to be a Delaporte distribution with parameters $\lambda = y_0\eta(t)$, $\alpha = b$ and $\beta = \frac{\eta(t)}{c}$. In the specific case where $y_0 = 0$, then N_t is negative binomial distributed. Assuming further that Y is exponentially distributed ($b = 1$), we get the geometric distribution.

EXAMPLE 3.4.– Let Y be inverse Gaussian distributed with p.d.f.

$$f_Y(y) = \sqrt{\frac{\beta}{2\pi\, y^3}} \exp\left(-\frac{1}{2}\,\frac{\beta(y-\mu)^2}{\mu^2\, y}\right) \mathbf{1}_{\{y>0\}}.$$

From [3.2], we get

$$p_k(t) = \int_0^\infty \frac{(y\eta(t))^k}{k!}\sqrt{\frac{\beta}{2\pi\, y^3}}\ \exp\left(-\frac{1}{2}\,\frac{\beta(y-\mu)^2 + 2y^2\mu^2\eta(t)}{\mu^2\, y}\right) dy.$$

The term in the exponential function can be written as

$$-\frac{\beta\,(y\sqrt{1+2\eta(t)\mu^2/\beta}-\mu)^2}{2\,\mu^2\, y} - \frac{\beta}{\mu}(\sqrt{1+2\eta(t)\mu^2/\beta}-1).$$

Therefore, we get

$$p_k(t) = \exp\left(-\frac{\beta}{\mu}(\sqrt{1+2\eta(t)\mu^2/\beta}-1)\right)\frac{\eta(t)^k}{k!\,\sqrt{(1+2\eta(t)\mu^2/\beta)^k}} \cdot \mathbb{E}(W^k),$$

where W is an inverse Gaussian distributed random variable with parameters β and (mean) $\mu/\sqrt{1+2\eta(t)\mu^2/\beta}$. Using formulas [1.13] for moments of the inverse Gaussian distribution given in Chapter 1, we obtain

$$p_k(t) = \exp\left(-\frac{\beta}{\mu}(\sqrt{1+2\eta(t)\mu^2/\beta}-1)\right)\left(\frac{\mu\,\eta(t)}{\sqrt{1+2\eta(t)\mu^2/\beta}}\right)^k \frac{1}{k!}\cdot$$
$$\sum_{u=0}^{k-1}\frac{(k-1+u)!}{(k-1-u)!\,u!}\left(\frac{\mu}{2\beta\sqrt{1+2\eta(t)\mu^2/\beta}}\right)^u. \qquad [3.6]$$

Frequently, the expected number of shocks up to any time t and other moments of N_t are of interest. According to [3.2], we can express the k-th ordinary and central moments of N_t as linear combinations of moments of Y weighted by powers of the deterministic function $\eta(t)$. An explicit expression for the ordinary moment is provided in theorem 3.8 just below in terms of Stirling numbers of the second kind [ABR 72]. We recall that these numbers are defined by their generating function

$$n^k = \sum_{u=1}^{k} S(k,u)\,n\,(n-1)\cdot\ldots\cdot(n-u+1) \quad k \le n, \qquad [3.7]$$

and obey the recurrence relation

$S(k,u) = S(k-1,u-1) + u\cdot S(k-1,u)$ for k > 0 with initial conditions

$S(0,0) := 1$, and $S(k,0) = S(0,k) = 0$.

THEOREM 3.5.– If the k-th moment of N_t exists, then

$$\mathbb{E}(N_t^k) = \sum_{u=1}^{k} S(k,u)\cdot\mathbb{E}(Y^u)\cdot\eta(t)^u \qquad [3.8]$$

where $S(k,u)$ are the Stirling numbers of the second kind.

PROOF.– We have to calculate

$$\mathbb{E}(N_t^k) = \sum_{n=0}^{\infty} n^k\mathbb{P}(N_t = k) = \sum_{n=0}^{\infty} n^k\mathbb{E}\left(\frac{(Y\eta(t))^k}{k!}\exp(-Y\eta(t))\right)$$
$$= \int_0^\infty \sum_{n=0}^{\infty} n^k\,\frac{(y\eta(t))^n}{n!}e^{-y\eta(t)}\mathrm{d}F_Y(y).$$

Based on [3.7], we get

$$\mathbb{E}(N_t^k) = \sum_{u=1}^{k} S(k,u) \int_0^\infty \sum_{n=0}^{\infty} n(n-1)\dots(n-u+1)$$
$$\times \frac{(y\eta(t))^n}{n!} e^{-y\eta(t)} \mathrm{d}F_Y(y).$$

The sum under the integral now is the u-th factorial moment of a Poisson distributed random variable with mean $y\,\eta(t)$ and it is $(y\,\eta(t))^u$. Therefore,

$$\mathbb{E}(N_t^k) = \sum_{u=1}^{k} S(k,u) \int_0^\infty (y\,\eta(t))^u \mathrm{d}F_Y(y) = \sum_{u=1}^{k} S(k,u) \cdot (\eta(t))^u \cdot \mathbb{E}(Y^u).$$

□

Remembering that the first values of Stirling numbers are

$$S(1,1) = 1, S(2,1) = 1, S(2,2) = 1, S(3,1) = 1, S(3,2) = 3, S(3,3) = 1,$$

the first three moments can now found to be

$$\begin{aligned} \mathbb{E}(N_t) =& \mathbb{E}(Y)\,\eta(t) \\ \mu_2(N_t) =& \mathrm{Var}(N_t) = \mathbb{E}(Y)\,\eta(t) + \mathrm{Var}(Y)\,\eta(t)^2 \\ \mu_3(N_t) =& \mathbb{E}(Y)\,\eta(t) + 3\,\mathrm{Var}(Y)\,\eta(t)^2 + \mu_3(Y)\,\eta(t)^3 \end{aligned} \quad [3.9]$$

where $\mu_k(\cdot)$ denotes the k-th central moment of a random variable. These results can be applied for parameter estimation through the moments method, which will be discussed in section 3.2.3.

Using the probabilities $p_k(t)$ provided in [3.2], we can also calculate the c.d.f. F_{T_n} and p.d.f. f_{T_n} of times of shocks T_n, $n \in \mathbb{N}^*$:

$$F_{T_n}(t) = \mathbb{P}(T_n \leq t) = \mathbb{P}(N_t \geq n) = 1 - \sum_{k=0}^{n-1} \mathbb{P}(N_t = k)$$

For the density, we get

$$f_{T_n}(t) = -\sum_{k=0}^{n-1} p_k'(t) = -\sum_{k=0}^{n-1} \frac{\mathrm{d}}{\mathrm{d}t} \int_0^\infty \frac{(y\,\eta(t))^k}{k!} e^{-y\,\eta(t)} \mathrm{d}F_Y(y)$$

$$= -\sum_{k=1}^{n-1} \int_0^\infty y\,\xi(t) \frac{(y\,\eta(t))^{k-1}}{(k-1)!} e^{-y\,\eta(t)} \mathrm{d}F_Y(y)$$

$$+ \sum_{k=0}^{n-1} \int_0^\infty y\,\xi(t) \frac{(y\,\eta(t))^{k}}{k!} e^{-y\,\eta(t)} \mathrm{d}F_Y(y)$$

$$= \int_0^\infty y\,\xi(t) \frac{(y\,\eta(t))^{n-1}}{(n-1)!} e^{-y\,\eta(t)} \mathrm{d}F_Y(y)$$

which gives

$$f_{T_n}(t) = n\,\frac{\xi(t)}{\eta(t)} \int_0^\infty \frac{(y\,\eta(t))^n}{n!} e^{-y\,\eta(t)} \mathrm{d}F_Y(y) = n\,\frac{\xi(t)}{\eta(t)} \mathbb{P}(N_t = n). \quad [3.10]$$

3.1.2. *Marking of the sequence* (T_n)

Let us now consider a marking of the sequence (T_n) [LAS 95]. At every time point T_n, a shock causes a random degradation. We describe the degradation increment at T_n by the mark Z_n. Let $(\mathbb{R}_+, \mathcal{B}(\mathbb{R}_+))$ be the space of marks and let G be a stochastic kernel $G : \mathbb{R}_+ \times \mathcal{B}(\mathbb{R}_+) \to \mathbb{R}_+$.

DEFINITION 3.6.– *The sequence* $\mathbf{N} = ((T_n, Z_n))$ *is said to be a* position-dependent G-marking *of* (T_n) *if* $Z_1, Z_2, \ldots$ *are conditionally independent given* (T_n), *and if, for all* $B \in \mathcal{B}(\mathbb{R}_+)$ *and all* $n \in \mathbb{N}$, *the following relation holds true:*

$$\mathbb{P}(Z_n \in B \,|\, (T_n)) = G(T_n, B) \qquad \mathbb{P} - a.s. \quad on \ (T_n < \infty). \quad [3.11]$$

Moreover, we assume that each mark Z_n and Y is conditionally independent given (T_n), i.e. $\mathbb{P}(Z_n \in B \,|\, (T_n), Y) = \mathbb{P}(Z_n \in B \,|\, (T_n))$. With a position-dependent marking, it is possible to describe a degradation model where the increment at the n-th shock depends on the time T_n of the n-th shock. Two examples are now provided.

EXAMPLE 3.7.– Let $t_0 \geq 0$ be a fixed time and let (U_n), (V_n) be two sequences of i.i.d. random variables with c.d.f. F_U and F_V, respectively. The sequence of degradation increments (Z_n) is defined by

$$Z_n := \mathbf{1}_{\{T_n \leq t_0\}} U_n + \mathbf{1}_{\{T_n > t_0\}} V_n$$

and $G(t, [0, x])$ is given by

$$G(t, [0, x]) = \mathbf{1}_{\{t \leq t_0\}} F_U(x) + \mathbf{1}_{\{t > t_0\}} F_V(x). \qquad [3.12]$$

That means that the distribution of degradation increments is changing at time t_0. For $t_0 = 0$, we get the independent marking.

EXAMPLE 3.8.– Let (U_n) be a sequence of non-negative i.i.d. random variables with c.d.f. F_U and let $\delta \in \mathbb{R}$. We assume that the sequence (U_n) is independent of (T_n). The sequence (Z_n) is defined by $Z_n = U_n \cdot e^{\delta T_n}$. The degradation increments can hence tend to be increasing ($\delta > 0$) or decreasing ($\delta < 0$). The stochastic kernel G is given by

$$G(t, [0, x]) = F_U(xe^{-\delta t}) \quad \text{for all } t, x \geq 0.$$

For $\delta = 0$, we have the special case of independent marking. In this case, G defines a probability measure that is independent of the time t.

3.1.3. *The sequence* (Z_n) *and the degradation process* (X_t)

If we consider a position-dependent marking of a doubly stochastic Poisson process where the filtration is given by $\mathcal{F}_t = \mathcal{F}_t^{\mathbf{N}} \vee \sigma(Y)$ (i.e. the frailty variable Y is observable), then by [LAS 95], the $(\mathbb{P}, \mathcal{F}_t)$ stochastic intensity kernel of $\mathbf{N} = ((T_n, Z_n))$ is

$$\lambda(t, B) = Y \xi(t) G(t, B) \quad \text{for all } B \in \mathcal{B}(\mathbb{R}_+) \text{ and all } t \geq 0,$$

where $\xi(t) = \eta'(t)$, $G(T_n, B) = \mathbb{P}(Z_n \in B \mid (T_n))$.

In the case of discrete random variables Z_n, we have

$$G(T_n, B) = \sum_{x \in B} G(T_n, \{x\})$$

and in the continuous case, we introduce the conditional density $g(t, x) = \frac{\partial}{\partial x} G(t, [0, x])$ of the marks Z_n, $n \in \mathbb{N}^*$ given $T_n = t$.

REMARK 3.9.– In the specific case where $Y = y_0 \ \mathbb{P} - a.s.$, let us recall that (T_n) are the points of a non-homogeneous Poisson process with deterministic intensity function $y_0 \cdot \xi(t)$. Alternatively, they can be seen as the points of a

Poisson random measure with intensity measure $y_0 \cdot \xi(t)\, dt$ (see proposition 2.18). Based on [ÇIN 11, theorem 3.2 p. 264], the sequence (T_n, Z_n) then appears as the points of a Poisson random measure M^{y_0} with intensity measure $y_0 \cdot \xi(t)\, dt \cdot G(t, dx)$. (Note that for an independent marking, it is a direct consequence of corollary 2.22). In the case of a general distribution for Y, given $Y = y_0$, the sequence (T_n, Z_n) is now conditionally identically distributed as the points of the same Poisson random measure M^{y_0}. Finally, in the specific case of an independent marking with $Y = y_0\ \mathbb{P} - a.s.$ and $\eta(t) = \lambda t$, the cumulative degradation process $(X_t)_{t \geq 0}$ can be seen to be a compound Poisson process and, consequently, it is a Lévy process.

Based on [3.11], the distribution of the n-th mark Z_n is given by

$$\mathbb{P}(Z_n \in B) = \mathbb{E}\left(G(T_n, B)\right) = \int_0^\infty G(t, B)\, f_{T_n}(t) \mathrm{d}t, \qquad [3.13]$$

where the density f_{T_n} of T_n has the representation $f_{T_n}(t) = n \cdot \frac{\xi(t)}{\eta(t)} \mathbb{P}(N_t = n), n \geq 1$ (see equation [3.10]). As we can see from equation [3.13], it is often difficult or strenuous to determine the distribution or the expectation of Z_n. Frequently, we are more interested in the expected cumulative degradation at any time t, which we now provide.

THEOREM 3.10.– Let $\mathbf{N} = ((T_n, Z_n))$ be a position-dependent marking of a doubly stochastic Poisson process with the stochastic intensity kernel

$$\lambda(t, B) = Y \cdot \xi(t)\, G(t, B), \quad B \in \mathcal{B}(\mathbb{R}_+).$$

Then, the Laplace transform of the cumulative degradation level at time t is provided by

$$\mathcal{L}_{X_t}(s) = \mathbb{E}\left(\exp\left(-Y \int_0^t \xi(u) \left(\int_{\mathbb{R}^+} \left(1 - e^{-sx}\right)\, G\left(u, \mathrm{d}x\right)\right) \mathrm{d}u\right)\right) [3.14]$$

for all $t, s \geq 0$.

Also, the mean cumulative degradation level at time t is given by

$$\mathbb{E}(X_t) = \mathbb{E}(Y) \int_0^t \xi(u) \cdot \int_0^\infty x \cdot G(u, \mathrm{d}x)\, \mathrm{d}u \qquad \text{for all } t \in \mathbb{R}_+, \qquad [3.15]$$

where, in case of a continuous marking, $G(u, \mathrm{d}x) = g(u, x)\mathrm{d}x$ with $g(t, x) = \frac{\partial}{\partial x} G(t, [0, x])$.

PROOF.– Let us first write

$$\mathcal{L}_{X_t}(s) = \mathbb{E}\left(e^{-sX_t}\right) = \mathbb{E}\left(h_{s,t}(Y)\right) \qquad [3.16]$$

where $h_{s,t}(y_0) = \mathbb{E}\left(e^{-sX_t}|Y = y_0\right)$. Using the Poisson random measure M^{y_0} with intensity measure $y_0 \cdot \xi(u)\, \mathrm{d}u \cdot g(u,x)\mathrm{d}x$ introduced in remark 3.9 and setting $f_{s,t}(u,x) = sx\, \mathbf{1}_{[0,t]}(u)$, we have

$$h_{s,t}(y_0) = \mathbb{E}\left(e^{-\sum_{n=1}^{+\infty} f_{s,t}(T_n,Z_n)}\right) = \mathbb{E}\left(e^{-M^{y_0} f_{s,t}}\right).$$

Theorem 2.21 now provides

$$\begin{aligned} h_{s,t}(y_0) &= \exp\left(-\iint_{\mathbb{R}+2} \left(1 - e^{-f_{s,t}(u,x)}\right) y_0 \cdot \xi(u)\, \mathrm{d}u \cdot g(u,x)\mathrm{d}x\right) \\ &= \exp\left(-y_0 \int_0^t \left(\int_{\mathbb{R}+} \left(1 - e^{-sx}\right) g(u,x)\mathrm{d}x\right) \xi(u)\, \mathrm{d}u\right) \end{aligned}$$

from where we easily derive [3.14] based on [3.16]. Equation [3.15] is a direct consequence using that $\mathbb{E}(X_t) = -\mathcal{L}'_{X_t}(0^+)$. □

Let us mention that higher order moments can easily be derived from the Laplace transform of X_t. Also, in the special case of an independent marking where the function $g(t,s)$ does not depend on t, we get the well-known result

$$\mathbb{E}(X_t) = \mathbb{E}(Y) \int_0^t \xi(s)\mathbb{E}(Z)\mathrm{d}s = \mathbb{E}(N_t) \cdot \mathbb{E}(Z).$$

Generally, the expectation $\mathbb{E}(X_t)$ depends on both the shape of the deterministic function η and the kernel G. Figure 3.2 shows the mean cumulative degradation process $(\mathbb{E}(X_t))_{t\geq 0}$ for the special position-dependent marking $Z_n = U_n\, e^{\delta T_n}$ from example 3.8. In this case $g(s,x) = e^{-\delta s}\, f_U(x\, e^{-\delta s})$ and the following relation holds true

$$\begin{aligned} \mathbb{E}(X_t) =& \mathbb{E}(Y) \int_0^t \xi(s) \cdot \int_0^\infty x \cdot e^{-\delta s} f_U(x\, e^{-\delta s})\mathrm{d}x\, \mathrm{d}s \\ =& \mathbb{E}(Y) \int_0^t \xi(x) e^{\delta s} \cdot \int_0^\infty x \cdot f_U(v)\mathrm{d}v\, \mathrm{d}s \\ =& \mathbb{E}(Y) \cdot \mathbb{E}(U_1) \int_0^t \xi(s) \cdot e^{\delta s}\mathrm{d}s. \end{aligned}$$

For the figure, we considered the function $\xi(t) = (\alpha + 1)t^{\alpha}$ from the Weibull type for different α. We selected the values $\delta = -0.2$ (decreasing degradation increments, left figure), $\delta = 0$ (independent marking) and $\delta = 0.2$ (increasing degradation increments, right figure).

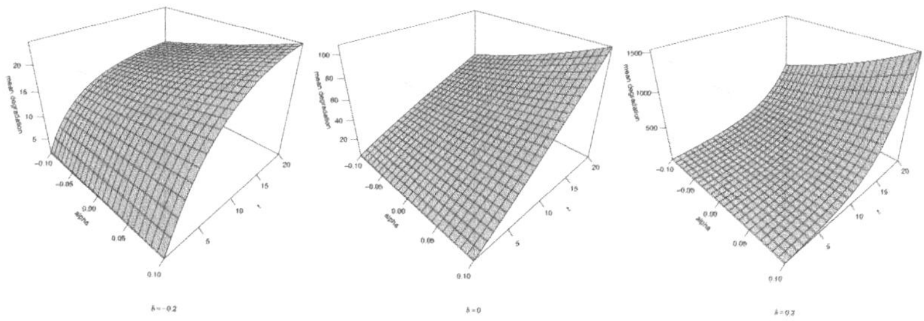

Figure 3.2. *Mean cumulative degradation process for different markings of* (T_n)

The c.d.f. of the cumulative degradation X_t at time t can be calculated by

$$\mathbb{P}(X_t \le x) = \mathbb{P}(N_t = 0)\mathbf{1}_{\{0 \le x\}} + \sum_{n=1}^{\infty} \mathbb{P}(Z_1 + \ldots + Z_n \le x, N_t = n).$$

In the case of independent marking, we get the simpler equation

$$\mathbb{P}(X_t \le x) = \mathbb{P}(N_t = 0)\mathbf{1}_{\{0 \le x\}} + \sum_{n=1}^{\infty} (F * (\mathrm{d}F)^{*(n-1)})(x) \cdot \mathbb{P}(N_t = n), \qquad [3.17]$$

where F is the c.d.f. of a mark and $(\mathrm{d}F)^{*(n-1)}$ represents the $(n-1)$-th convolution of the corresponding probability measure $\mathrm{d}F$. In both cases, the c.d.f. of X_t can be numerically computed through numerical Laplace transform inversion, as provided by most classical softwares (with more or less success however). Finding a closed form and computable expression is a much more complicated problem, even in the case of independent marking. Methods for calculating the degradation distribution for special distributions of marks can be found in [SCH 95] and [GRA 97]. Explicit expressions for this distribution are mostly known only for simple models. One of them is provided in the following example.

EXAMPLE 3.11.– An independent G-marking of (T_n) is considered where the marks are exponentially distributed with parameter $p > 0$. Equation [3.14] provides:

$$\mathcal{L}_{X_t}(s) = \mathbb{E}\left(\exp\left(-Y \int_0^t \xi(u) \left(\int_{\mathbb{R}+} \left(1 - e^{-sx}\right) pe^{-px} \mathrm{d}x\right) \mathrm{d}u\right)\right)$$
$$= \mathbb{E}\left(e^{-Y\,\eta(t)\frac{s}{p+s}}\right).$$

Assuming furthermore that Y is also exponentially distributed with parameter c, we get:

$$\mathcal{L}_{X_t}(s) = \int_{\mathbb{R}+} ce^{-cy} e^{-y\,\eta(t)\frac{s}{p+s}} \mathrm{d}y = \frac{c}{c + \eta(t)\frac{s}{p+s}}.$$

A partial fraction expansion now provides

$$\mathcal{L}_{X_t}(s) = \frac{c}{c + \eta(t)} + \frac{\eta(t)}{c + \eta(t)} \frac{\frac{cp}{c+\eta(t)}}{\frac{cp}{c+\eta(t)} + s}.$$

Remembering that $s \to 1$ (respectively, $s \to \frac{\lambda}{\lambda+s}$) is the Laplace transform of the Dirac mass δ_0 at point 0 (respectively, of an exponential distribution ν_λ with parameter λ), the previous expression can be readily inverted, which shows that the distribution of X_t is a mixture between the Dirac mass δ_0 and ν_λ with $\lambda = \frac{cp}{c+\eta(t)}$:

$$\mathbb{P}_{X_t}(\mathrm{d}u) = \frac{c}{c + \eta(t)} \delta_0(\mathrm{d}u) + \frac{\eta(t)}{c + \eta(t)} \frac{cp}{c + \eta(t)} e^{-\frac{cp}{c+\eta(t)}u} \mathrm{d}u$$

(so that $\mathbb{P}(X_t = 0) = \frac{c}{c+\eta(t)}$). We easily derive the c.d.f. of X_t, with

$$F_{X_t}(x) = \int_{[0,x]} \mathbb{P}_{X_t}(\mathrm{d}u) = 1 - \frac{\eta(t)}{c + \eta(t)} \cdot \exp\left(-\frac{cpx}{c + \eta(t)}\right) \quad [3.18]$$

for all $x \geq 0$.

3.1.4. *The first passage time* τ_h

The system is here assumed to start from an initial deterioration level X_0 at time $T_0 := 0$, so that the accumulated deterioration at time t is

$$X_0 + \sum_{n=1}^{\infty} Z_n \mathbf{1}_{\{T_n \le t\}}.$$

For any $h > 0$, the first passage time τ_h above h is

$$\tau_h = \inf \left\{ t : X_0 + \sum_{n=1}^{\infty} Z_n \mathbf{1}_{\{T_n \le t\}} \ge h \right).$$

If $X_0 = x_0$ is deterministic and if it is possible to calculate the degradation distribution, then it is easy to derive the distribution of the first passage time through

$$\mathbb{P}(\tau_h > t) = \mathbb{P}(X_t < h - x_0), \qquad [3.19]$$

where $X_t = \sum_{n=1}^{\infty} Z_n \mathbf{1}_{\{T_n \le t\}}$.

EXAMPLE 3.12.– Let us again consider the assumptions of example 3.11. We choose the fixed parameters $c = 4$ and $h - x_0 = 100$. Figure 3.3 shows the corresponding densities of the first passage time τ_h for the different deterministic functions (from left to right)

$$\eta(t) = t^{\alpha+1}, \; \eta(t) = t \exp(0.03 t^{\alpha+0.2}) \text{ and } \eta(t) = t(1 + \log(1 + 1.85 t^{\alpha})).$$

The parameter α is varying and the number of shocks up to time t tends to increase with increasing α.

Now we consider a random initial state X_0, which is assumed to be independent of the marked point process.

In case of an independent marking, we get

$$\mathbb{P}(\tau_h > t) = \mathbb{P}(X_0 \le h)\, \mathbb{P}(N_t = 0) + \sum_{n=1}^{\infty} \mathbb{P}(X_0 + Z_1 + \ldots + Z_n \le h) \cdot \mathbb{P}(N_t = n). \qquad [3.20]$$

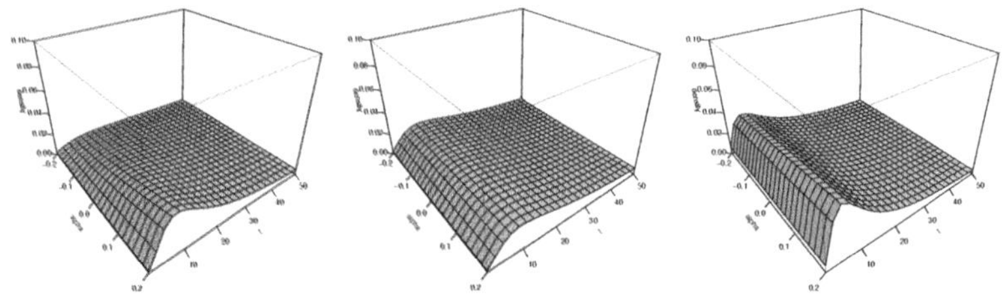

Figure 3.3. *Density of the first passage time τ_h for different η*

To use this formula, we have to determine the distribution of $X_0 + Z_1 + \ldots + Z_n$, where in general the distribution of X_0 does not coincide with that of any mark Z_i, $i \geq 1$.

Setting $X_t = \sum_{n=1}^{\infty} Z_n \mathbf{1}_{\{T_n \leq t\}}$, an alternate method is to note that

$$\mathbb{P}(\, \tau_h > t\,) = \mathbb{P}(\, X_t \leq h - X_0\,) = \int_{[0,h]} F_{X_t}(h - x_0)\, \mathrm{d}F_{X_0}(x_0). \qquad [3.21]$$

EXAMPLE 3.13.– In addition to the assumptions of example 3.11, let X_0 be an exponentially distributed random variable with parameter p_0 (independent of the marked point process). Starting from [3.21] and [3.18], we have:

$$\begin{aligned}
&\mathbb{P}(\tau_h > t) \\
&= \int_0^h \left(1 - \frac{\eta(t)}{c + \eta(t)} \cdot \exp\left(-\frac{cp\,(h - x_0)}{c + \eta(t)}\right)\right) p_0\, e^{-p_0 x_0}\, \mathrm{d}x_0 \\
&= 1 - e^{-p_0\, h} - \frac{p_0\, \eta(t)}{c + \eta(t)} \exp\left(-\frac{c\,p\,h}{c + \eta(t)}\right) \int_0^h \\
&\quad \times \exp\left(-\left(p_0 - \frac{cp}{c + \eta(t)}\right) x_0\right)\, \mathrm{d}x_0.
\end{aligned}$$

In case $\eta(t) \neq c(p - p_0)/p_0$, we get

$$\begin{aligned}
\mathbb{P}(\tau_h > t) = 1 &- \frac{1}{p_0\, \eta(t) + c\,(p_0 - p)} \\
&\times \left(p_0\, \eta(t) \exp\left(-\frac{c\,p\,h}{c + \eta(t)}\right) + c\,(p_0 - p)\, e^{-p_0 \cdot h}\right)
\end{aligned}$$

after straightforward computations.

In the specific case where $\eta(t) = c(p - p_0)/p_0$, we have:

$$\mathbb{P}(\tau_h > t) = 1 - e^{-p_0 \cdot h}\left(1 + \frac{h\, p_0(p - p_0)}{p}\right).$$

Also, the probability that the system fails at time 0 is

$$\mathbb{P}(\tau_h = 0) = \mathbb{P}(X_0 > h) = e^{-h\, p_0}.$$

In Figure 3.4, the influence of the distribution of the initial state X_0 on the distribution of the first passage time τ_h is represented. We considered example 3.13, where we chose an exponentially distributed random variable Y with parameter $c = 4$, a function η of logistic type $\eta(t) = t(1 + \log(1 + 0.8t^{0.4}))$, and a degradation level $h = 25$. The parameter p_0 of the distribution of X_0 is varying: $0.05 \leq p_0 \leq 1.05$. With increasing $\mathbb{E}(X_0)$, i.e. with decreasing p_0, the probability of a failure at time $t = 0$ is increasing. As noticed previously, it can happen that the system fails at time $t = 0$, which can be seen in Figure 3.4 for small p_0.

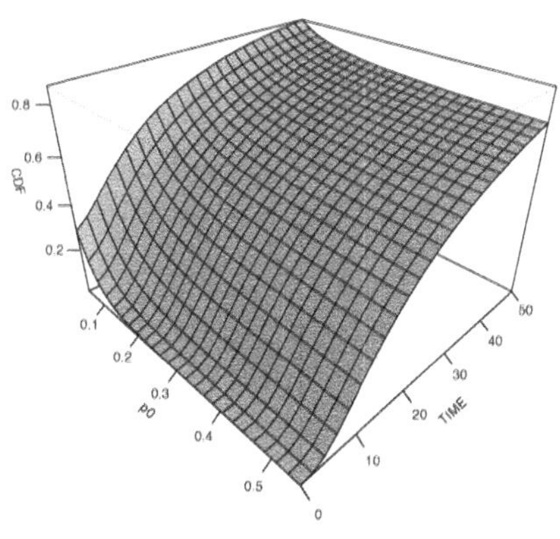

Figure 3.4. *Distribution function of τ_h for different X_0*

In general, it is difficult to calculate the distribution of failure time and its moments. It is even possible that the expectation of the first passage time does not exist. Let us consider the case that the process starts in 0 at time 0 ($X_0 = 0$). Since $\tau_h \geq T_1$, it follows that $\mathbb{E}(\tau_h) = \infty$ as soon as $\mathbb{E}(T_1) = \infty$.

Based on [3.10], [3.2] and on Fubini's theorem, we have:

$$\begin{aligned}\mathbb{E}(T_1) &= \int_0^\infty t\frac{\xi(t)}{\eta(t)}\mathbb{E}\left(Y\eta(t)e^{-Y\eta(t)}\right)\mathrm{d}t \\ &= \mathbb{E}\left(\int_0^\infty t\,\xi(t)Ye^{-Y\eta(t)}\mathrm{d}t\right).\end{aligned}$$

Writing $t = \int_0^t \mathrm{d}u$ and using Fubini's theorem again, we get:

$$\mathbb{E}(T_1) = \mathbb{E}\left(\int_0^\infty \left(\int_u^\infty \xi(t)Ye^{-Y\eta(t)}\mathrm{d}t\right)\mathrm{d}u\right).$$

Assuming $\lim_{t\to\infty}\eta(t) = \infty$, we derive:

$$\mathbb{E}(T_1) = \mathbb{E}\left(\int_0^\infty e^{-Y\eta(u)}\mathrm{d}u\right) = \int_0^\infty\int_0^\infty e^{-y\,\eta(t)}\mathrm{d}F_Y(y)\,\mathrm{d}t. \tag{3.22}$$

Based on this expression, we can see that $\mathbb{E}(T_1)$ depends on the distribution of Y as well as of $\eta(t)$ and can be infinite, leading to an infinite mean time to failure. This is illustrated in the following example for a deterministic function from the Weibull type.

EXAMPLE 3.14.– Let Y be uniformly distributed on $[0, b]$ with $0 < b$ and $\eta(t) = t^{\alpha+1}$ with $\alpha > -1$. Then, we get by substituting $t^{\alpha+1} = z$

$$\begin{aligned}\mathbb{E}(T_1) &= \int_0^\infty \frac{1-e^{-b\eta(t)}}{b\eta(t)}\,\mathrm{d}t = \int_0^\infty \frac{1-e^{-bt^{\alpha+1}}}{bt^{\alpha+1}}\,\mathrm{d}t \\ &= \int_0^\infty \frac{1-e^{-bz}}{b(\alpha+1)z^{1+\alpha/(\alpha+1)}}\,\mathrm{d}z = \infty \quad \text{for} \quad \alpha \leq 0.\end{aligned}$$

If Y is $\mathcal{G}(b, c)$-distributed, then

$$\begin{aligned}\mathbb{E}(T_1) &= \int_0^\infty \left(\frac{c}{c+\eta(t)}\right)^b \mathrm{d}t \\ &= \int_0^\infty \left(\frac{c}{c+t^{\alpha+1}}\right)^b \mathrm{d}t = \infty \text{ for } b\cdot(\alpha+1) \leq 1.\end{aligned}$$

3.2. Statistical inference

We now come to the problem of estimating the parameters of the model, which is necessary for practical applications. We here consider both likelihood and moments methods. We assume the existence of a family of probability measures $\mathbb{P}_\Theta = (\mathbb{P}_\theta, \theta \in \Theta)$ on $(\Omega, \mathcal{F})$. Let θ be given as $\theta = (\theta^Y, \theta^T, \theta^Z) \in \mathbb{R}^p$ with $p = u + v + w$, where $\theta^Y \in \mathbb{R}^u$ is a parameter of the distribution function F_Y of Y, $\theta^T \in \mathbb{R}^v$ a parameter of the deterministic term η (or ξ) and $\theta^Z \in \mathbb{R}^w$ a parameter of the kernel G, respectively.

3.2.1. *Maximum likelihood estimations*

First, we have to determine the likelihood function for a marked point process and to begin with, we need to introduce its conditional intensity function given the available information. We start by considering a point process $(N_t)_{t\geq 0}$ with points $(T_n)_{n\geq 1}$ (no double points) and $T_0 = 0$. For all $k \geq 1$, let $f_k(\cdot|T_1, \cdots, T_{k-1})$ and $F_k(\cdot|T_1, \cdots, T_{k-1})$ be the respective p.d.f. and c.d.f. of the conditional distribution of T_k given $T_1, \cdots, T_{k-1}$. Also, let $h_k(t|T_1, \cdots, T_{k-1})$ be the corresponding hazard rate function with

$$h_k(t|T_1, \cdots, T_{k-1}) = \frac{f_k(t|T_1, \cdots, T_{k-1})}{\bar{F}_k(t|T_1, \cdots, T_{k-1})} \qquad [3.23]$$

for all $t > T_{k-1}$, which can be readily seen to be equivalent to

$$f_k(t|T_1, \cdots, T_{k-1}) = h_k(t|T_1, \cdots, T_{k-1}) e^{-\int_{T_{k-1}}^t h_k(s|T_1, \cdots, T_{k-1}) ds} \qquad [3.24]$$

$$\bar{F}_k(t|T_1, \cdots, T_{k-1}) = e^{-\int_{T_{k-1}}^t h_k(s|T_1, \cdots, T_{k-1}) ds} \qquad [3.25]$$

for all $t > T_{k-1}$.

Following [DAL 03, p. 231], the conditional intensity function (or hazard function) of $(N_t)_{t\geq 0}$ is now defined piecewise through

$$\bar{\lambda}(t) = \begin{cases} h_1(t) \text{ if } 0 < t \leq T_1, \\ h_k(t|T_1, \cdots, T_{k-1}) \text{ if } T_{k-1} < t \leq T_k \text{ and } k \geq 2, \end{cases}$$

or equivalently through

$$\bar{\lambda}(t) = h_{N_{t-}+1}(t|T_1, \cdots, T_{N_{t-}}).$$

Based on this last formula, we can see that $\lambda(t)$ is a random variable that depends on the history of the process, namely on the $\sigma-$field $\mathcal{F}_{t^-}$ generated by the internal history of the point process up to time t^- (namely on $[0, t)$). Remembering [3.23], the conditional intensity function (or hazard function) of $(N_t)_{t\geq 0}$ can also be heuristically interpreted in the following way, setting $\mathbb{P}_{\mathcal{F}_{t^-}}$ to be the conditional probability given $\mathcal{F}_{t^-}$:

$$\begin{aligned}\bar{\lambda}(t)\mathrm{d}t &= \frac{\mathbb{P}_{\mathcal{F}_{t^-}}\left(T_{N_{t^-}+1} \in (t, t+\mathrm{d}t]\right)}{\mathbb{P}_{\mathcal{F}_{t^-}}\left(T_{N_{t^-}+1} > t\right)} \\ &= \mathbb{P}_{\mathcal{F}_{t^-}}\left(T_{N_{t^-}+1} \in (t, t+\mathrm{d}t] | T_{N_{t^-}+1} > t\right) \\ &= \mathbb{P}_{\mathcal{F}_{t^-}}[N(\mathrm{d}t) \geq 1]\end{aligned}$$

where $N(\mathrm{d}t)$ stands for the number of points in the infinitesimal interval $(t, t+\mathrm{d}t]$.

Based on the assumption that no points coincide, the number of points $\mathrm{d}N(t)$ in $(t, t+\mathrm{d}t]$ can only be either zero or one, resulting in a Bernoulli trial. This provides:

$$\bar{\lambda}(t)\mathrm{d}t = \mathbb{P}_{\mathcal{F}_{t^-}}[\mathrm{d}N(t) = 1] = \mathbb{E}_{\mathcal{F}_{t^-}}[\mathrm{d}N(t)] = \mathbb{E}[\mathrm{d}N(t)|\mathcal{F}_{t^-}]$$

and the intensity function can be seen as a measure of the "conditional risk of the occurrence of an event at t, given the realization of the process over the interval $[0, t)$" [DAL 03, p. 232].

Let us now assume that a trajectory of (T_n) is observed on $[0, t]$ and let $\mathbf{t} = (t_1, \ldots, t_{n_t})$ be the observed data (where n_t stands for an observation of N_t). Based on [3.24] and [3.25] for the second and third lines just below, the likelihood function may be written as:

$$\begin{aligned}L_1(\theta; \mathbf{t}, \mathbf{x}) &= f_1(t_1) \times \prod_{k=2}^{n_t} f_k(t_k|t_1, \cdots, t_{k-1}) \times F_{n_t+1}(t|t_1, \cdots, t_{n_t}) \\ &= h_1(t_1)\, e^{-\int_0^{t_1} h_1(s)\mathrm{d}s} \\ &\quad \times \prod_{k=2}^{n_t} h_k(t_k|t_1, \cdots, t_{k-1}) e^{-\int_{t_{k-1}}^{t_k} h_k(s|t_1, \cdots, t_{k-1})\mathrm{d}s}\end{aligned}$$

$$\times e^{-\int_{t_{n_t}}^{t} h_{n_t+1}(s|t_1,\cdots,t_{n_t})\mathrm{d}s}$$

$$= \bar{\lambda}(t_1)\, e^{-\int_0^{t_1}\bar{\lambda}(s)\mathrm{d}s} \times \prod_{k=2}^{n_t} \bar{\lambda}(t_k) e^{-\int_{t_{k-1}}^{t_k}\bar{\lambda}(s)\mathrm{d}s} \times e^{-\int_{t_{n_t}}^{t}\bar{\lambda}(s)\mathrm{d}s}$$

$$= \prod_{k=1}^{n_t} \bar{\lambda}(t_k) \times e^{-\int_0^t \bar{\lambda}(s)\mathrm{d}s}. \qquad [3.26]$$

For a marked point process $\mathbf{N} = ((T_n, Z_n))$, the conditional intensity function is now defined as

$$\lambda^*(t,x) = \bar{\lambda}(t) g(t,x),$$

where $\bar{\lambda}(t)$ is the conditional intensity function of the point process (T_n), and $g(t,x)$ is the conditional density of marks given t (assuming continuous marks). The conditional intensity function $\lambda^*(t,x)$ can be interpreted in a similar way as for a point process through the following:

$$\lambda^*(t,x)\, \mathrm{d}t\, \mathrm{d}x = \mathbb{E}[N(\mathrm{d}t \times \mathrm{d}x)|\mathcal{F}_{t^-}],$$

where $N(\mathrm{d}t \times \mathrm{d}x)$ stands for the number of points in the infinitesimal interval $(t, t + \mathrm{d}t]$ with mark in the infinitesimal interval $[x, x + \mathrm{d}x)$.

Let us now assume that a trajectory of $\mathbf{N} = ((T_n, Z_n))$ is observed on $[0,t]$ and let $(\mathbf{t}, \mathbf{x}) = \{(t_1, \ldots, t_{n_t}), (x_1, \ldots, x_{n_t})\}$ be the observed data. The likelihood function may be written as:

$$L_3(\theta; \mathbf{t}, \mathbf{x}) = L_1(\theta; \mathbf{t}, \mathbf{x}) \times L_2(\theta; \mathbf{t}, \mathbf{x}) \qquad [3.27]$$

where $L_1(\theta; \mathbf{t}, \mathbf{x})$ corresponds to the likelihood of the point process (T_n) given by [3.26] and where $L_2(\theta; \mathbf{t}, \mathbf{x})$ corresponds to the product of the conditional p.d.f.s of the marks x_i's given the arrival times t_i's: $L_2(\theta; \mathbf{t}, \mathbf{x}) = \prod_{k=1}^{n_t} g(t_k, x_k)$. This provides

$$L_3(\theta; \mathbf{t}, \mathbf{x}) = \prod_{k=1}^{n_t} \bar{\lambda}(t_k) g(t_k, x_k) \times e^{-\int_0^t \bar{\lambda}(s)\mathrm{d}s}. \qquad [3.28]$$

Let us now consider the case of a position-dependent marking of a doubly stochastic Poisson process $\mathbf{N} = ((T_n, Z_n))$.

We first consider the case where the frailty variable Y is observable, namely the case where the filtration is given by $\mathcal{F}_t = \mathcal{F}_t^{\mathbf{N}} \vee \sigma(Y)$, where we recall that $\mathcal{F}_t^{\mathbf{N}}$ refers to the internal filtration of $\mathbf{N}$. Then by [LAS 95], for continuous distributions of marks, the conditional intensity function of $\mathbf{N} = ((T_n, Z_n))$ given $\mathcal{F}_{t-}$ is

$$\lambda(t, x) = Y\,\xi(t)\,g(t, x),$$

which is also called $(\mathbb{P}, \mathcal{F}_t)$ stochastic intensity of $\mathbf{N}$.

Assuming that a trajectory of $((T_n, Z_n))$ is fully observed on some given time interval $[0, t]$ (jump times + marks) as well as the frailty variable Y, we can write down the likelihood function as a product of three terms, where each of them contains one of the parameters $\theta^Y, \theta^T, \theta^Z$. Consequently, they can be estimated independently one from the other. This case hence is easy to deal with in a classical way. However, it is not realistic, because the frailty variable Y corresponds to an individual characteristic, which has no reason to be observable. We consequently assume throughout the following that *the frailty variable Y is not observed.*

In some situations, it is possible that a component failure is the result of a degradation process that cannot be observed. Namely, it is possible that only failures are observed. In that case, we can write down the likelihood function based on the distribution of the first passage time, which contains all the parameters. By the maximum likelihood method, it is hence possible to estimate all the parameters in the model. It should however be noted that this can usually be done only through numerical optimization procedures, even for simple examples. Also, the sample size must be large to estimate all parameters. We do not pursue on this case in the following.

In other situations, the degradation process can be observed up to a given horizon time t, which means that each time of shock is observed up to time t, together with the corresponding increment of degradation (but not the frailty variable). Note that this is a different situation from the case where some observations are stopped at a failure time τ_h smaller than t. Just as in Chapter 1, this last situation is complicated by the fact that the censoring time τ_h contains some information about the process parameters, leading to so-called "informative censoring". This case is not dealt with in the following, where we concentrate on the previously quoted case of a degradation process observed up to a given horizon time t.

To be more specific, we hereafter assume that a trajectory of $\mathbf{N} = ((T_n, Z_n))$ is observed on $[0,t]$ and we let $(\mathbf{t}, \mathbf{x}) = \{(t_1, \dots, t_{n_t}), (x_1, \dots, x_{n_t})\}$ to be the observed data. The observed history is hence given by $(\mathcal{F}_t^{\mathbf{N}})$, where we recall that the frailty variable Y is not observable. As assumed in section 3.1.2, each mark Z_n and Y is conditionally independent given (T_n), so that Y has an influence only on the distribution of (T_n). Let $\bar{\lambda}^Y = \xi(t)\, Y$ be the conditional intensity function of (T_n) given all the information $(\mathcal{F}_t)$ (including Y) and let $\bar{\lambda}$ be the conditional intensity function of (T_n) given $(\mathcal{F}_t^{\mathbf{N}})$. As $\mathcal{F}_t^{\mathbf{N}} \subset \mathcal{F}_t$, we know from the innovation theorem [AND 93, page 80] that

$$\bar{\lambda}(t) = \mathbb{E}\left(\bar{\lambda}^Y | \mathcal{F}_{t^-}^{\mathbf{N}}\right) = \xi(t)\, \mathbb{E}\left(Y | \mathcal{F}_{t^-}^{\mathbf{N}}\right).$$

Now, for each $n \in \mathbb{N}$, we have:

$$\begin{aligned} \mathbb{E}\left(Y | N_{t^-} = n\right) &= \frac{\mathbb{E}\left(Y\, \mathbf{1}_{\{N_{t^-}=n\}}\right)}{\mathbb{P}\left(N_{t^-} = n\right)} = \frac{\mathbb{E}\left(Y\, \mathbb{E}\left(\mathbf{1}_{\{N_{t^-}=n\}} | Y\right)\right)}{\mathbb{P}\left(N_{t^-} = n\right)} \\ &= \frac{\mathbb{E}\left(Y^{n+1} e^{-Y\, \eta(t)}\right)}{\mathbb{E}\left(Y^n e^{-Y\, \eta(t)}\right)} \end{aligned}$$

where, for the last equality, we use the fact that given Y, the r.v. N_{t^-} is almost surely Poisson distributed with parameter $Y\, \eta\,(t)$ and [3.10]. This provides

$$\bar{\lambda}(t) = \xi(t; \theta^T) \frac{\int_0^\infty y^{N_{t^-}+1} e^{-y\, \eta(t)} \mathrm{d}F_Y(y; \theta^Y)}{\int_0^\infty y^{N_{t^-}} e^{-y\, \eta(t)} \mathrm{d}F_Y(y; \theta^Y)}. \qquad [3.29]$$

Now we are able to present the likelihood function that is of the shape $L_1 \times L_2$ as in [3.27], where L_2 is just the same as in [3.27] and where L_1 is computed with $\bar{\lambda}(t)$ provided by [3.29].

Setting $T_0 = 0$, let us first notice that for any $k = 0, 1, \dots$

$$N_{s^-} = k - 1 \quad \text{for all } s \in (T_{k-1}, T_k] \quad \text{and} \quad N_{s^-} = N_t \quad \text{for all } s \in (T_{N_t}, t].$$

Furthermore, for any $k = 0, 1, \ldots$, we also have:

$$\int_{T_{k-1}}^{T_k} \bar{\lambda}(s)\mathrm{d}s = \int_{T_{k-1}}^{T_k} \xi(s;\theta^T)\,\frac{\int_0^\infty y^k e^{-y\,\eta(s;\theta^T)}\mathrm{d}F_Y(y;\theta^Y)}{\int_0^\infty y^{k-1} e^{-y\,\eta(s;\theta^T)}\mathrm{d}F_Y(y;\theta^Y)}\mathrm{d}s$$
$$= -\int_{T_{k-1}}^{T_k} \frac{\mathrm{d}}{\mathrm{d}s}\log\Big(\int_0^\infty y^{k-1} e^{-y\,\eta(s;\theta^T)}\mathrm{d}F_Y(y;\theta^Y)\Big)\mathrm{d}s$$
$$= -\log\left(\int_0^\infty y^{k-1} e^{-y\,\eta(s;\theta^T)}\mathrm{d}F_Y(y;\theta^Y)\right)\Bigg|_{s=T_{k-1}}^{s=T_k}$$

We get:

$$\int_0^t \bar{\lambda}(s)\mathrm{d}s = \sum_{k=1}^{N_t}\int_{T_{k-1}}^{T_k}\bar{\lambda}(s)\mathrm{d}s + \int_{T_k}^t \bar{\lambda}(s)\mathrm{d}s$$
$$= \sum_{k=1}^{N_t-1}\log(\int_0^\infty y^k\, e^{-y\,\eta(T_k;\theta^T)}\mathrm{d}F_Y(y;\theta^Y))$$
$$-\sum_{k=1}^{N_t}\log(\int_0^\infty y^{k-1}\, e^{-y\,\eta(T_k;\theta^T)}\mathrm{d}F_Y(y;\theta^Y))$$
$$-\log\Big(\int_0^\infty y^{N_t}\, e^{-y\,\eta(t;\theta^T)}\mathrm{d}F_Y(y;\theta^Y)\Big).$$

The log-likelihood function based on the observation of one deterioration process is now given by

$$\log L(t;\theta) = \sum_{k=1}^{N_t}\log g(T_k, Z_k;\theta^Z) + \sum_{k=1}^{N_t}\log\xi(T_k;\theta^T)$$
$$+\sum_{k=1}^{N_t}\log\Big(\frac{\int_0^\infty y^k e^{-y\,\eta(T_k;\theta^T)}\mathrm{d}F_Y(y;\theta^Y)}{\int_0^\infty y^{k-1} e^{-y\,\eta(T_k;\theta^T)}\mathrm{d}F_Y(y;\theta^Y)}\Big) - \int_0^t \bar{\lambda}(s)\mathrm{d}s$$
$$= \sum_{k=1}^{N_t}\log g(T_k, Z_k;\theta^Z) + \sum_{k=1}^{N_t}\log\xi(T_k;\theta^T)$$
$$+ \log\Big(\int_0^\infty y^{N_t} e^{-y\,\eta(t;\theta^T)}\mathrm{d}F_Y(y;\theta^Y)\Big).$$

Now, let us assume that we observe $m \geq 1$ independent copies of the process $\mathbf{N}$ on $[0, t]$, and let T_n^i, Z_n^i and N_t^i refer to the i-th copy

$(i = 1, \ldots, m)$. Our aim is to determine an maximum likelihood estimator estimator of θ based on these m copies. The maximum (partial) likelihood equations $\frac{\partial}{\partial \theta_r} \log(L(t;\theta)) = 0$ are given by

$$\frac{\partial \log L(t;\theta)}{\partial \theta_j^Y} = \sum_{i=1}^{m} \frac{\partial}{\partial \theta_j^Y} \log \left(\int_0^\infty y^{N_t^i} \exp(-y\,\eta(t;\theta^T)) \mathrm{d}F_Y(y;\theta^Y) \right) \tag{3.30}$$

$$\begin{aligned}\frac{\partial \log L(t;\theta)}{\partial \theta_k^T} = &\sum_{i=1}^{m} \left(\sum_{n=1}^{N_t^i} \frac{\partial \log(\xi(T_n^i;\theta^T))}{\partial \theta_k^T} \right.\\ &\left. + \frac{\partial}{\partial \theta_k^T} \log \left(\int_0^\infty y^{N_t^i} \exp(-y\,\eta(t;\theta^T)) \mathrm{d}F_Y(y;\theta^Y) \right) \right)\end{aligned} \tag{3.31}$$

$$\frac{\partial \log L(t;\theta)}{\partial \theta_l^Z} = \sum_{i=1}^{m} \sum_{n=1}^{N_t^i} \frac{\partial}{\partial \theta_l^Z} \log \left(g(T_n^i, Z_n^i; \theta^Z) \right) \tag{3.32}$$

provided the derivatives exist. In general, these equations can only be solved by numerical methods. It can be seen that equation [3.32] depends only on estimator $\hat{\theta}^Z$ so that this equation can be solved separately. This is illustrated in the following example.

EXAMPLE 3.15.– We consider the position-dependent marking $Z_n = U_n \cdot e^{\delta T_n}$ from example 3.8, where the i.i.d. random variables U_n are exponentially distributed with p.d.f. $f_U(u) = p\,e^{-p\,u} \mathbf{1}_{\{u \geq 0\}}$. The parameter to estimate hence is $\theta^Z = (\delta, p)$. Here, we get for $n = 1, \ldots, N_t,\ i = 1, \ldots, m$

$$\begin{aligned}\log \left(g(T_n^i, Z_n^i; \theta^Z) \right) &= \log \left(f_U(Z_n^i \cdot e^{-\delta\, T_n^i}) \right) - \delta T_n^i \\ &= \log p - p Z_n^i \cdot e^{-\delta\, T_n^i} - \delta T_n^i .\end{aligned}$$

It is easy to find that $\hat{p}$ is given as

$$\hat{p} = \frac{\sum_{i=1}^{m} N_t^i}{\sum_{i=1}^{m} \sum_{n=1}^{N_t} Z_n^i \exp(-\hat{\delta}\, T_n^i)},$$

where $\hat{\delta}$ is the solution of a nonlinear equation

$$h(\hat{\delta}) = (\sum_{i=1}^{m} N_t^i) \frac{\sum_{i=1}^{m} \sum_{n=1}^{N_t^i} T_n^i \cdot Z_n^i\, e^{-\hat{\delta}\, T_n^i}}{\sum_{i=1}^{m} \sum_{n=1}^{N_t^i} Z_n^i\, e^{-\hat{\delta}\, T_n^i}} - \sum_{i=1}^{m} \sum_{n=1}^{N_t^i} T_n^i = 0 .$$

It can be shown that this equation has a unique solution, based on the fact that $h(\hat{\delta})$ is continuous and decreasing and $\lim_{\hat{\delta}\to-\infty} h(\hat{\delta}) > 0$ and $\lim_{\hat{\delta}\to\infty} h(\hat{\delta}) < 0$.

Now, we consider equations [3.30] and [3.31] to find estimators for θ^Y and θ^T.

Recalling the probabilities of numbers of failures provided by [3.2], we can note that

$$\log\left(\int_0^\infty y^{N_t}\exp(-y\,\eta(t;\theta^T))\mathrm{d}F_Y(y;\theta^Y)\right) = \log\left(\frac{N_t!}{(\eta(t;\theta^T))^{N_t}}\,p_{N_t}(t)\right)$$
$$= \log(N_t!) - N_t\log(\eta(t;\theta^T)) + \log(p_{N_t}(t))\,. \qquad [3.33]$$

Equation [3.30] now becomes

$$\frac{\partial \log L(t;\theta)}{\partial\theta_j^Y} = \sum_{i=1}^m \frac{\partial}{\partial\theta_j^Y}\log\left(p_{N_t^i}(t)\right) = 0,$$

which can be solved to express $\hat{\theta}^Y$ with respect of $\hat{\theta}^T$. We now provide an example for this method.

EXAMPLE 3.16.– Let Y be $\mathcal{G}(b,c)$-distributed so that $\theta^Y = (b,c)$. Then, we get from [3.5] with $y_0 = 0$

$$p_{N_t^i}(t) = \frac{\Gamma(b+N_t^i)}{\Gamma(b)N_t^i!}\left(\frac{c}{c+\eta(t;\theta^T)}\right)^b\left(\frac{\eta(t;\theta^T)}{c+\eta(t;\theta^T)}\right)^{N_t^i}$$
$$= \frac{\prod_{l=0}^{N_t^i-1}(b+l)}{N_t^i!}\,\frac{c^b(\eta(t;\theta^T))^{N_t^i}}{(c+\eta(t;\theta^T))^{b+N_t^i}} \quad \text{and}$$

$$\sum_{i=1}^m \log(p_{N_t^i}(t)) = \sum_{i=1}^m\sum_{l=0}^{N_t^i-1}\log(b+l) - \sum_{i=1}^m\log(N_t^i!) + mb\log(c)$$
$$+\log(\eta(t;\theta^T))\sum_{i=1}^m N_t^i - \log(c+\eta(t;\theta^T))\sum_{i=1}^m(b+N_t^i)\,.$$

Taking the derivatives with respect to c and b, it is easy to find that

$$\hat{c} = \frac{\hat{b}\,\eta(t;\hat{\theta}^T)}{\bar{N}_t} \quad\text{with}\quad \bar{N}_t = \frac{1}{m}\sum_{i=1}^m N_t^i$$

where $\hat{b}$ is the solution of

$$\log\left(\frac{\hat{b}+\bar{N}_t}{\hat{b}}\right) = \frac{1}{m}\sum_{i=1}^{m}\sum_{l=0}^{N_t^i-1}\frac{1}{\hat{b}+l}.$$

Differentiating [3.33] with respect to θ^T and inserting the estimator for c, we get

$$\sum_{i=1}^{m}\frac{\partial}{\partial\theta_k^T}\log\left(\int_0^\infty y^{N_t^i}\exp(-y\,\eta(t;\theta^T))\mathrm{d}F_Y(y;\theta^Y)\right)$$
$$= -\frac{\partial\eta(t;\theta^T)}{\partial\theta_k^T}\frac{\sum_{i=1}^{m}(b+N_t^i)}{c+\eta(t;\theta^T)} = -\frac{\partial\log(\eta(t;\theta^T))}{\partial\theta_k^T}\sum_{i=1}^{m}N_t^i.$$

As a result, the parameter θ^T can be estimated independently of θ^Y.

REMARK 3.17.– It can be shown that our restriction to the uncensored case always leads to a simplification of [3.31] as in the previous example and that the following condition is always satisfied $\mathbb{P}_\theta - a.s.$:

$$\sum_{i=1}^{m}\frac{\partial}{\partial\theta_k^T}\log\left(\int_0^\infty y^{N_t^i}\exp(-y\,\eta(t;\theta^T))\mathrm{d}F_Y(y;\theta^Y)\right)$$
$$= -\frac{\partial\log(\eta(t;\theta^T))}{\partial\theta_k^T}\sum_{i=1}^{m}N_t^i. \quad [3.34]$$

For more details, we refer to [WEN 99].

An immediate consequence is the possibility to determine the estimator $\hat{\theta}^T$ of the system [3.31] without knowing the distribution of the random variable Y. Actually, if we take into consideration relation [3.34], then the system of equations for $\hat{\theta}^T$ is given by

$$\sum_{i=1}^{m}\left(\sum_{n=1}^{N_t^i}\frac{\partial}{\partial\theta_k^T}\log(\xi(T_n^i;\theta^T)) - N_t^i\cdot\frac{\partial}{\partial\theta_k^T}\log(\eta(t;\theta^T))\right)\Bigg|_{\theta^T=\hat{\theta}^T} = 0, \quad [3.35]$$

$k = 1,\ldots,v$. Thus, for any distribution function F_Y, the same estimator $\hat{\theta}^T$ is obtained as under the assumption of a non-homogeneous Poisson process. Having determined $\hat{\theta}^T$, we can next solve the system of equations [3.30] for θ^Y.

EXAMPLE 3.18.– Let us estimate the parameters of the deterministic part in the doubly stochastic Poisson process. We consider an intensity of the Weibull type. With $\eta(t;\alpha) = t^{\alpha+1}$, $\alpha > -1$ and $\xi(t;\alpha) = (\alpha+1)\, t^{\alpha}$, equation [3.35] provides

$$\frac{\partial}{\partial\alpha}\log(\xi(t;\alpha)) = \frac{1}{\alpha+1} + \log(t), \qquad \frac{\partial}{\partial\alpha}\log\eta(t;\alpha) = \log(t)$$

and

$$\hat{\alpha} = \frac{\sum_{i=1}^{m} N_t^i}{\log(t)\cdot\sum_{i=1}^{m} N_t^i - \sum_{i=1}^{m}\sum_{n=1}^{N_t^i}\log(T_n^i)} - 1 . \qquad [3.36]$$

For both other intensities (log linear and logistic), equation [3.35] leads to a system of nonlinear equations that should be solved numerically.

3.2.2. *Confidence estimations*

In this section, we deal with the large sample properties of the maximum likelihood estimator $\hat{\theta} = (\hat{\theta}^Y, \hat{\theta}^T, \hat{\theta}^Z)$ based on observations up to time $t < \infty$. It is used that the score vector process, i.e. the vector of the first derivatives of the log-likelihood process $(\log L(t;\theta))$, with respect to $\theta \in \mathbb{R}^p$ is a local square integrable $(\mathbb{P}_\theta, \mathcal{F}_t^{\mathrm{N}})$ martingale. Under mild differentiation and continuity requirements on η, ξ, g as well as on F_Y with respect to θ, Taylor series expansions on the score vector process are valid. And further conditions ensure a proper behavior of the remaining terms in these Taylor expansions.

According to [AND 93] (for (θ^Y, θ^T)) and Wendt [WEN 99] (for θ^Z), it follows that under the regularity conditions with a probability tending to one the likelihood equations have a consistent solution $\hat{\theta} = (\hat{\theta}^Y, \hat{\theta}^T, \hat{\theta}^Z)$. Furthermore, $\hat{\theta}$ is asymptotically multinormally distributed around the true value θ_0. The asymptotic covariance matrix term $(\sigma_{j,l}(\theta_0))^{-1}$ may be estimated consistently by $\left(-\frac{1}{m}\left.\frac{\partial^2 \log L(\tau;\theta)}{\partial\theta_j\partial\theta_l}\right|_{\theta=\hat{\theta}}\right)^{-1}$.

These results can be used to calculate confidence regions and acceptance areas for tests on the parameter vector θ applying the well-known Wald's test statistic W and the likelihood ratio test statistic R, respectively. Both are asymptotically χ^2 distributed with r degrees of freedom, where r is the number of unknown parameters. They are given by

$$
\begin{aligned}
W &= (\hat{\theta} - \theta_0)^T \, I(t; \theta_0) \, (\hat{\theta} - \theta_0) \\
R &= 2 \left(\ln(L(t; \hat{\theta})) - \ln(L(t; \theta_0)) \right)
\end{aligned}
$$

In Wald's statistic, the information matrix I can be replaced by a consistent estimator $\hat{I}(\theta) = I(\hat{\theta})$. The advantage of the likelihood ratio statistic is that it is not necessary to calculate the information matrix.

Let us give some examples for such confidence regions.

EXAMPLE 3.19.– We consider the doubly stochastic Poisson process from examples 3.15, 3.16 and 3.18. As we have seen, the parameters θ^T and θ^Z can be estimated independently of the other parameters, while the estimator of θ^Y depends on $\hat{\theta}^T$.

First, let us find a confidence interval for parameter α of a Weibull intensity. The estimated information is $(\hat{\alpha} + 1)^2 / \sum_{i=1}^m N_t^i$. Therefore, a $(1 - \gamma)$ confidence interval for α based on the Wald statistic is given by

$$
-z_{1-\gamma/2} \leq \frac{(\hat{\alpha} - \alpha)}{(\hat{\alpha} + 1)} \sqrt{\sum_{i=1}^{m} N_t^i} \leq z_{1-\gamma/2}
$$

where $z_{1-\gamma/2}$ is the quantile of order $1 - \gamma/2$ of the standard Gaussian distribution. From the likelihood ratio statistic, we get the equation

$$
2\left(m\bar{N}_t \log\left(\frac{1+\hat{\alpha}}{1+\alpha}\right) + (\hat{\alpha} - \alpha)\left(\sum_{i=1}^{m} \sum_{n=1}^{N_t^i} \log(T_n^i) - m\bar{N}_t \log(t) \right) \right) \leq z_{1-\gamma/2}^2
$$

Now, we consider the position-dependent marking of (T_n) from example 3.15. Here, we have two unknown parameters. With estimated information, we get the Wald statistic

$$
\begin{aligned}
&\left(\frac{\hat{p} - p}{\hat{p}}\right)^2 \sum_{i=1}^{m} N_t^i - 2 \sum_{i=1}^{m} \sum_{n=1}^{N_t^i} T_n^i Z_n^i e^{-\hat{\delta} T_n^i} (\hat{p} - p)(\hat{\delta} - \delta) \\
&\quad + \hat{p} \sum_{i=1}^{m} \sum_{n=1}^{N_t^i} (T_n^i)^2 Z_n^i e^{-\hat{\delta} T_n^i} (\hat{\delta} - \delta)^2 \leq -2 \log(\gamma) \,.
\end{aligned}
$$

The likelihood ratio statistic is given by

$$\sum_{i=1}^{m}\sum_{n=1}^{N_t^i}\left\{\log\left(\frac{\hat{p}}{p}\right) - Z_n^i\left(\hat{p}\,e^{-\hat{\delta}T_n^i} - pe^{-\delta T_n^i}\right) - T_n^i(\hat{\delta}-\delta)\right\} \leq -\log(\gamma).$$

Last, joint confidence intervals for the parameters (b, c) of the distribution of Y are given by

$$(\hat{c}-c)^2\left(\frac{m\hat{b}}{\hat{c}^2} - \frac{m(\hat{b}+\bar{N}_t)}{(\hat{c}+\eta(t;\hat{\theta}^T))^2}\right) - 2\frac{m\eta(t;\hat{\theta}^T)}{\hat{c}(\hat{c}+\eta(t;\hat{\theta}^T))}(\hat{c}-c)(\hat{b}-b)$$

$$+ (\hat{b}-b)^2\sum_{i=1}^{m}\sum_{l=0}^{N_t^i-1}\frac{1}{(\hat{b}+l)^2} \leq -2\log(\gamma) \qquad \text{(Wald statistic)} \qquad \text{and}$$

$$\sum_{i=1}^{m}\sum_{l=0}^{N_t^i-1}\log\left(\frac{\hat{b}+l}{b+l}\right) + m\hat{b}\log\left(\frac{\hat{c}}{\hat{c}+\eta(t;\hat{\theta}^T)}\right) - mb\log\left(\frac{c}{c+\eta(t;\hat{\theta}^T)}\right)$$

$$- m\bar{N}_t\log\left(\frac{\hat{c}+\eta(t;\hat{\theta}^T)}{c+\eta(t;\hat{\theta}^T)}\right) \leq -\log(\gamma) \qquad \text{(likelihood ratio statistic).}$$

Next, we show the results for a simulated data file. For this, 20 processes were simulated up to time $t = 3$ with parameters:

– $c = 1$ and $b = 1.5$ for the gamma frailty;

– $\alpha = 1$ for the deterministic part of the intensity (Weibull intensity with shape 2);

– $p = 2$ and $\delta = 0.3$ for the distribution of marks.

The realized values of frailty are varying in the data file from 0.28 (smallest) to 4.69 (largest). Within the sample, the number of jumps varies from 3 to 41. The overall number of jumps is 359. We got the following parameter estimates:

– $\hat{\alpha} = 0.943$ with 95% confidence intervals [0.742;1.144] (Wald statistic) and [0.749;1.151] (likelihood ratio statistic);

– $\hat{\delta} = 0.321$ and $\hat{p} = 2.081$ for the parameters in the distribution of marks;

– $\hat{b} = 2.378$ and $\hat{c} = 1.120$ for the parameters in the frailty distribution.

In Figure 3.5, joint confidence regions are plotted for the parameters (p, δ) (left) and (b, c) (right). The star represents the point estimate and $+$ is the true parameter. We get relatively good estimates for α, δ and p. The estimators of the parameters of the frailty distribution, however, have a large variance. This is not surprising, because they were estimated from only 20 realizations and they depend on the estimator of α.

3.2.3. *Moment estimates*

Let us consider again the case of an observable counting process and unobservable frailty variable Y. Let us further assume that the deterministic function $\eta(t)$ is known and that we are only interested in estimating the parameters of the distribution of Y.

We denote by $m_k(N_t), k = 1, 2, 3$, the empirical (central) moments of N_t:

$$m_1(N_t) = \frac{1}{m} \sum_{i=1}^{m} N_t^i, \;\; m_k(N_t) = \frac{1}{m} \sum_{i=1}^{m} (N_t^i - m_1(N_t))^k, k = 2, 3. \quad [3.37]$$

According to [3.9], the first three (central) moments of N_t are provided by:

$$\begin{aligned} \mathbb{E}(N_t) =& \mathbb{E}_\theta(Y) \cdot \eta(t) \\ \mu_2(N_t) =& \mathbb{E}_\theta(Y) \cdot \eta(t) + \mu_2^\theta(Y) \cdot \eta(t)^2 \\ \mu_3(N_t) =& \mathbb{E}_\theta(Y) \cdot \eta(t) + 3\mu_2^\theta(Y) \cdot \eta(t)^2 + \mu_3^\theta(Y) \cdot \eta(t)^3. \end{aligned}$$

where $\mu_k(\cdot)$ denotes the k-th central moment of a random variable.

Based on the moments method, the central moments on the left-hand side are next matched with their empirical versions. We derive the following approximate expressions for the moments of Y with respect to the empirical moments of N_t and the function η:

$$\mathbb{E}_\theta(Y) = \eta(t)^{-1} \cdot m_1(N_t) \quad [3.38]$$

$$\mu_2^\theta(Y) = \eta(t)^{-2} \left(m_2(N_t) - m_1(N_t) \right) \quad [3.39]$$

$$\mu_3^\theta(Y) = \eta(t)^{-3} \left(m_3(N_t) - 3\, m_2(N_t) + 2\, m_1(N_t) \right). \quad [3.40]$$

Now, it is possible to find moment estimates for all parameters of the distribution of Y.

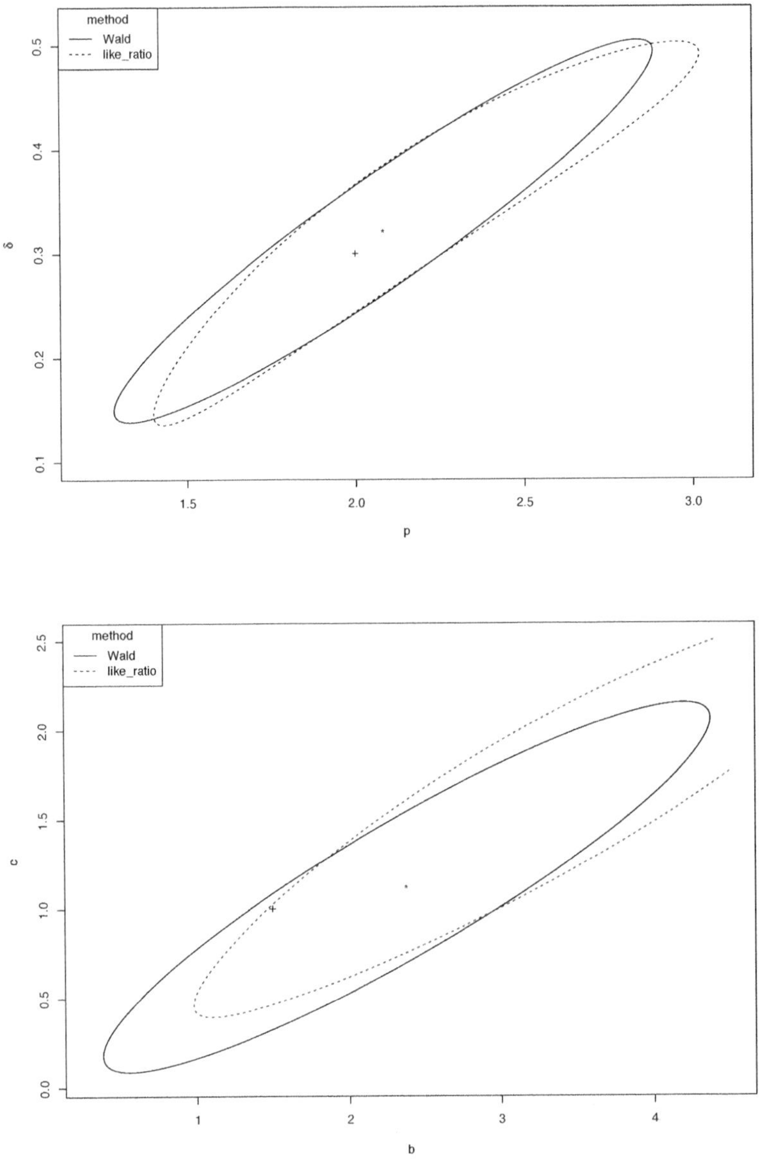

Figure 3.5. *Joint confidence estimates for the process parameters*

EXAMPLE 3.20.– Let us consider the three examples from section 3.1.1:

1) If Y is uniformly distributed in $[a, b]$, we get from [3.38] and [3.39] with $\mathbb{E}_\theta(Y) = (a+b)/2$, $\mu_2^\theta(Y) = (b-a)^2/12$, and taking into account $0 \leq a < b$

$$\hat{a} = \frac{m_1(N_t)}{\eta(t;\hat{\theta}^T)} - \sqrt{3\mathcal{D}^2}, \quad \hat{b} = \frac{m_1(N_t)}{\eta(t;\hat{\theta}^T)} + \sqrt{3\mathcal{D}^2}$$

with

$$\mathcal{D}^2 := \frac{m_2(N_t) - m_1(N_t)}{\eta(t;\hat{\theta}^T)^2}.$$

As a difference to the maximum likelihood method, a unique admissible estimator exists if $\hat{a} \geq 0$ and $\mathcal{D}^2 > 0$. The assumption $\mathcal{D}^2 > 0$ is fulfilled for sufficient large values of m because $\mathcal{D}^2$ is a consistent estimate of the variance $\mu_2^\theta(Y)$ of Y;

2) If $Y - y_0$ is gamma distributed, its first three moments are:

$$\mathbb{E}_\theta(Y) = \frac{b}{c} + y_0, \quad \mu_2^\theta(Y) = \frac{b}{c^2} \quad \text{and} \quad \mu_3^\theta(Y) = 2\,\frac{b}{c^3}.$$

From [3.38]–[3.40], we get the unique moment estimators

$$\hat{c} = 2\,\frac{m_2(N_t) - m_1(N_t)}{\mathcal{J}}\,\eta(t;\hat{\theta}^T), \quad \hat{y}_0 = \frac{m_1(N_t)}{\eta(t;\hat{\theta}^T)} - 2\,\frac{(m_2(N_t) - m_1(N_t))^2}{\mathcal{J}\cdot\eta(t;\hat{\theta}^T)}$$

$$\hat{b} = 4\,\frac{\Big(m_2(N_t) - m_1(N_t)\Big)^3}{\mathcal{J}^2} \quad \text{with } \mathcal{J} = m_3(N_t) - 3\,m_2(N_t) + 2\,m_1(N_t).$$

Note that MLE was considered only for $y_0 = 0$ due to the complexity to estimate three parameters through this method (see equation [3.5]).

3) For an inverse Gaussian distributed Y with $\mathbb{E}_\theta(Y) = \mu$ and $\mu_2^\theta(Y) = \mu^3/\beta$, equations [3.38] and [3.39] give the unique estimators

$$\hat{\mu} = \frac{m_1(N_t)}{\eta(t;\hat{\theta}^T)} \quad \text{and} \quad \hat{\beta} = \frac{m_1(N_t)^3}{\eta(t;\hat{\theta}^T)\cdot(m_2(N_t) - m_1(N_t))}.$$

For this distribution, we get the same $\hat{\mu}$ from the moment method as from the maximum likelihood method.

The advantages of moment estimators in comparison to maximum likelihood estimators are their simple form and the fact that they often can be found explicitly.

But it is well known that, in general, maximum likelihood estimates have better properties and that moment estimators are not asymptotically efficient, see [VOI 96]. In the following section, we numerically compare both bias and variance of the two methods as an example.

3.2.4. *Comparison of maximum likelihood and moment estimates*

Let $Y - y_0$ Gamma distributed and let η be Weibull with $\eta(t;\alpha) = t^{\alpha+1}$ and $\alpha > -1$. Several sizes of samples are considered, with $m = 50$, $m = 100$, $m = 250$ and $m = 500$. For each realization, a path of the degradation process was simulated with the true parameter $\theta^Y = (b, c, y_0) = (1.2, 2.4, 0.5)$ up to time $t = 10$. Three different values of the parameter α are envisioned in the deterministic part $\eta(t;\alpha)$. For $\alpha = -0.3$, the derivative of $\eta(t;\alpha)$ is a decreasing function. The expected number of jumps up to time $t = 10$ is 5.01. If $\alpha = 0$, we get a linear cumulative intensity (or a constant hazard function) and the expected number of jumps up to time $t = 10$ is 10. At last, for $\alpha = 0.3$ the derivative of $\eta(t;\alpha)$ is increasing and the expected number of jumps up to time $t = 10$ is 19.95. Such a simulation was repeated 750 times and from these 750 parameters estimates, the mean and the variance of the estimator were calculated. The results are shown in Table 3.1.

In case of an unknown parameter α in the deterministic part, it can be also estimated by

$$\hat{\alpha} = \frac{\sum_{i=1}^{m} N_t^i}{\log(t) \cdot \sum_{i=1}^{m} N_t^i - \sum_{i=1}^{m} \sum_{n=1}^{\mathbf{N}^i(t)} \log(T_n^i)} - 1,$$

(see equation [3.36]). $\hat{\alpha}$ is a maximum likelihood estimator that does not contain any other parameter. In Table 3.2, the results on the same simulations data are shown in Table 3.1, where α is now unknown and hence has to be estimated.

From the simulation results, we can make the following observations:

1) *influence of* α: in both cases, we can see that for $\alpha = 0.3$, the variances of the estimator $\hat{y}_0$ (in both cases, MLE and ME) are smaller than for $\alpha = 0$ or $\alpha = -0.3$. The variances of the moment estimators for $\hat{b}$ and $\hat{c}$ are also smaller for $\alpha = 0.3$ than for $\alpha = 0$ or $\alpha = -0.3$, while α does not influence much the variances of the maximum likelihood estimators;

2) the variances of the moment estimators are two to four times larger than the variances of the maximum likelihood estimators;

3) both bias and variance are clearly smaller if the parameter θ^T is known (except for the moment estimate $\hat{b}$, which is independent of η). The ratio between the variances of both maximum likelihood and moment estimators is however the same for known or unknown θ^T.

		$m = 50$		$m = 100$		$m = 250$		$m = 500$	
		Mean	Variance	Mean	Variance	Mean	Variance	Mean	Variance
$\alpha = -0.3$									
MLE	c	2.614	1.798	2.435	0.762	2.386	0.302	2.401	0.140
	y_0	0.599	0.018	0.561	0.016	0.520	0.008	0.508	0.004
	b	0.861	0.184	0.975	0.190	1.119	0.110	1.171	0.062
ME	c	2.910	3.902	2.824	2.208	2.826	1.475	2.758	0.893
	y_0	0.480	0.047	0.458	0.041	0.436	0.031	0.444	0.023
	b	1.680	2.552	1.708	2.026	1.754	1.507	1.650	1.052
$\alpha = 0.0$									
MLE	c	2.197	0.694	2.205	0.435	2.351	0.262	2.339	0.149
	y_0	0.574	0.013	0.560	0.010	0.532	0.005	0.536	0.003
	b	0.977	0.249	1.034	0.198	1.154	0.146	1.125	0.078
ME	c	3.189	2.533	2.846	1.325	2.771	0.876	2.613	0.427
	y_0	0.431	0.037	0.457	0.027	0.466	0.019	0.489	0.010
	b	1.982	2.030	1.716	1.257	1.621	0.855	1.431	0.375
$\alpha = 0.3$									
MLE	c	2.497	0.706	2.484	0.573	2.392	0.268	2.446	0.143
	y_0	0.521	0.010	0.519	0.007	0.523	0.004	0.497	0.002
	b	1.282	0.404	1.270	0.336	1.198	0.147	1.231	0.079
ME	c	3.361	2.066	3.133	1.399	2.758	0.672	2.602	0.347
	y_0	0.388	0.030	0.418	0.024	0.464	0.013	0.470	0.007
	b	2.276	2.110	2.004	1.457	1.596	0.625	1.417	0.279

Table 3.1. *Empirical moments of maximum likelihood (MLE) and moment (ME) estimators ($\theta_0^Y = (1.2, 2.4, 0.5)$)*

The moment estimates are easy to calculate, whereas in many cases, it is difficult to find maximum likelihood estimates. Moreover, there are even problems, in which the maximum likelihood estimate of the parameters does not exist.

		$m = 50$		$m = 100$		$m = 250$		$m = 500$	
		Mean	Variance	Mean	Variance	Mean	Variance	Mean	Variance
$\alpha = -0.3$									
MLE	c	2.187	3.217	2.000	0.913	2.125	0.481	2.197	0.407
	y_0	0.683	0.040	0.666	0.040	0.623	0.027	0.607	0.024
	b	0.818	0.157	0.891	0.141	1.026	0.084	1.072	0.054
	α	-0.328	0.011	-0.337	0.010	-0.337	0.009	-0.333	0.009
ME	c	2.801	3.942	2.729	3.067	2.641	1.622	2.607	1.144
	y_0	0.548	0.066	0.533	0.066	0.523	0.049	0.527	0.041
	b	1.621	2.179	1.739	2.357	1.669	1.410	1.572	0.916
$\alpha = 0.0$									
MLE	c	2.194	1.156	2.177	0.695	2.209	0.625	2.299	0.403
	y_0	0.594	0.029	0.594	0.028	0.602	0.028	0.592	0.019
	b	1.072	0.304	1.157	0.207	1.194	0.147	1.137	0.084
	α	-0.031	0.014	-0.040	0.014	-0.048	0.012	-0.036	0.010
ME	c	3.128	3.691	2.800	1.645	2.660	1.400	2.503	0.678
	y_0	0.438	0.051	0.471	0.041	0.516	0.038	0.532	0.025
	b	2.082	2.271	1.885	1.264	1.721	0.826	1.485	0.394
$\alpha = 0.3$									
MLE	c	2.571	0.924	2.652	1.001	2.394	0.495	2.363	0.286
	y_0	0.510	0.014	0.520	0.014	0.546	0.011	0.552	0.010
	b	1.424	0.516	1.455	0.378	1.269	0.148	1.160	0.059
	α	0.286	0.010	0.280	0.009	0.277	0.007	0.278	0.006
ME	c	3.324	2.478	3.234	1.748	2.687	0.900	2.494	0.497
	y_0	0.394	0.038	0.421	0.032	0.489	0.020	0.517	0.015
	b	2.297	1.926	2.171	1.535	1.639	0.703	1.409	0.250

Table 3.2. *Empirical moments of maximum likelihood (MLE) and moment (ME) estimators ($\theta_0^Y = (2.4, 0.5, 1.2)$)*

On the other hand, the maximum likelihood estimators have a noticeable smaller variance as moment estimators. The ratio of the variances of both maximum likelihood and moment estimators becomes smaller with increasing sample size.

4

Model Selection and Application to Real Data Sets

This chapter illustrates the use of the various deterioration models and statistical estimation procedures previously discussed in this book on a few real data sets. We however had no real data set with piecewise constant trajectories at our disposal so the doubly stochastic marked Poisson process is not considered here as a possible model, only the Wiener or gamma process. Apart from the statistical inference of the models, other practical issues are considered as follows: computation of remaining useful lifetimes, comparison between models, etc. To this end, the three data sets discussed in the Introduction are considered (for a precise description of these data, please refer to the Introduction).

4.1. Takeda device data

As observed in the Introduction, these data sets contain not only positive increments but also null and negative increments. As a result, the gamma process is not adapted to model the data. Also, as shown in Figure I.1, we can see that the data do not start from 0 at time 0. Consequently, we suggest to use a shifted Wiener process starting from some level x_0 at some time t_0, as proposed in the model described by [1.4]. Note that both x_0 and t_0 are unknown and hence need to be estimated, together with the parameters μ and σ^2 of the underlying Wiener process. Since the first observation times are identical for all units (in fact, observation times are the same for all units – see the description of the data set in Introduction), we can use the explicit expression of MLE given by equation [1.15]. The results are reported in

Table 4.1. The confidence intervals are computed by bootstrapping, using the Jackknife method.

t_0	x_0	μ	σ
3.9726	−0.6435	0.5003	0.0874
[3.3000, 4.7686]	[−0.9563; −0.3307]	[0.4830; 0.5116]	[0.0268; 0.0730]

Table 4.1. *Takeda device data – parameters estimation and 95% confidence intervals*

In Figure 4.1, we have plotted the estimated average degradation and the estimated quantiles at levels of 2.5% and 97.5%.

As specified in the Introduction, the critical threshold is assumed to be equal to $h = \log(0.15)$ of transconductance degradation [LU 97]. It follows that three of the five devices have not fail until the end of the experiment. For each of these devices, a random variable of interest is its conditional residual lifetime given all the available information, namely given the final degradation level reached at the end of the experiment. Setting τ_h to be the first passage time of the critical level h, given that $X_s = x$ (with $x < h$) and that the system has not failed before time s, the residual lifetime is identically distributed as $\tau_h - s$ given $\{\tau_h \geq s\} \cap \{X_s = x\}$. Setting $\tilde{\tau}_{h-x}$ to be the first passage time of the level $h - x$ by a standard Wiener process with similar parameters (μ, σ) (and $t_0 = x_0 = 0$), the survival function of the residual lifetime at time s given that $X_s = x$ (or at (s, x) for short) is now provided by

$$\begin{aligned} \mathrm{RL}_{s,x}(u) &= \mathbb{P}[\tau_h - s > u | \tau_h > s; X_s = x] \\ &= \mathbb{P}[\tilde{\tau}_{h-x} > u], \end{aligned}$$

for any $u \geq 0$, using the Markov property and the homogeneity of the standard Wiener process. Note that $\tilde{\tau}_{h-x}$ is inverse Gaussian distributed with parameters $\mu_1 = (h - x)/\mu$ and $\beta = (h - x)^2/\sigma^2$ whose c.d.f. is given as:

$$\begin{aligned} \forall t \geq 0, \; F_{\tilde{\tau}_{h-x}}(u) = {} & \Phi\left(\frac{\mu u - h + x}{\sigma\sqrt{u}}\right) \\ & + \exp\left(\frac{2\mu(h - x)}{\sigma^2}\right) \Phi\left(-\frac{\mu u + h - x}{\sigma\sqrt{u}}\right). \end{aligned}$$

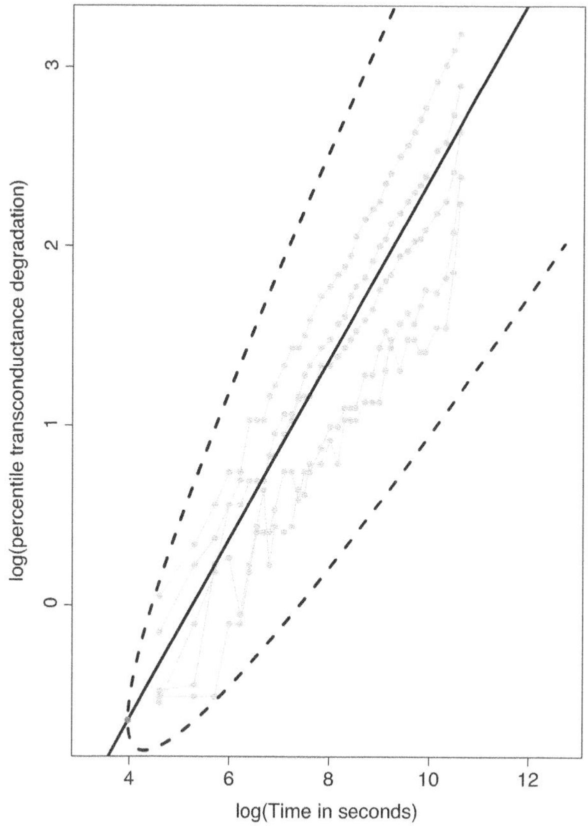

Figure 4.1. *Takeda device data: estimated average degradation and quantiles*

Consequently, the mean residual lifetime at (s, x) is equal to $(h - x)/\mu$. In Figure 4.2, we have plotted the estimated survival function of the residual lifetimes for the three unfailed devices (plugging the estimates obtained from the observation of all units). In this figure, we can observe the influence on the residual lifetime of the final level of each device at the end of the experiment.

4.2. Laser data

Based on Figure I.1, we can see that for the laser data, the degradation paths are monotone. Hence, we can fit both a standard Wiener process and a standard homogeneous gamma process (we have $t_0 = 0$ and $x_0 = 0$). A point of interest here is to compare the results provided by these two models. After having given the estimation of the parameters, we compare the estimated average degradation (and the estimated quantiles). Then, we consider the estimations of the probability density function of the time to failure. At last, we propose two approaches to choose a model from between the two models that have been fitted. The first approach is a simple graphical tool, which can however be used for any degradation model, whereas the second approach is based on statistical tests.

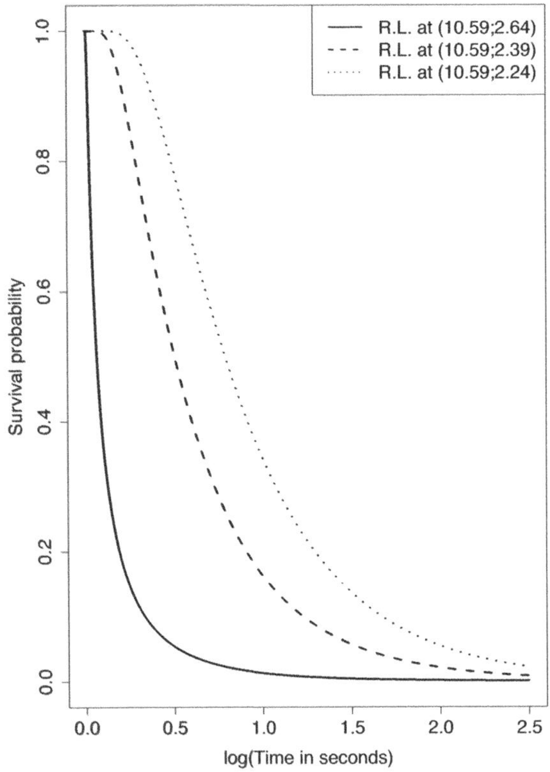

Figure 4.2. *Takeda device data: survival function of residual lifetimes for unfailed devices*

4.2.1. *Parameter estimation*

We first fit a homogeneous gamma process to the data set. We recall that items are observed at the same regular instants (balanced case). Parameters have been estimated both using the maximum likelihood method (see equations [2.45a] and [2.45b]) and using the moments method (see equation [2.46]). The results are reported in Table 4.2. Confidence intervals have been calculated by bootstrapping using the Jackknife method. Since the maximum likelihood method and the moments method lead to closed results, we will consider, in the following, only the MLE.

	a	b
MLE	0.0288	14.1183
	[0.0286; 0.0292]	[14.0550; 14.3798]
MME	0.0259	12.7162
	[0.0258, 0.0263]	[12.6612; 12.9776]

Table 4.2. *Laser data – parameters estimations for the gamma process and 95% confidence intervals*

We now fit a Wiener process to the laser data. We have computed the MLE of the parameters given in equation [1.19]. The results are reported in Table 4.3 with confidence intervals computed by bootstrapping (Jackknife method, here also).

μ	σ
0.0020372	0.0001602
[0.002028; 0.002042]	[0.000156; 0.000161]

Table 4.3. *Laser data – parameters estimation for the Wiener process and 95% confidence intervals*

4.2.2. *Comparison of the estimations for the average degradation and the time to failure*

In Figure 4.3, we have plotted the estimated average degradation and the (pointwise) estimated quantiles at levels of 2.5% and 97.5% for the two models. We can notice that there is no strong difference between the two estimations of these quantities and some curves are even nearly superimposed, in particular the average degradations. In both cases, a non-negligible number of observations are out of the (pointwise) dispersion intervals.

Let us now consider the estimation of the time-to-failure distributions still considering the two fitted models. In Figure 4.4, we have plotted the estimated probability density function of the time to failure for the two models. Here again, there is no strong difference between the two estimations. Such a result is not surprising and has already been reported by Tsai *et al.* [TSA 11].

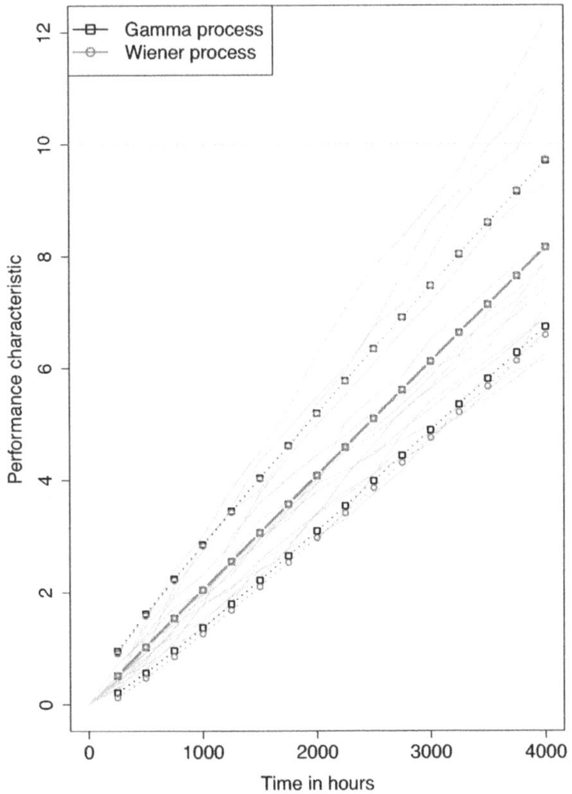

Figure 4.3. *Laser data*

4.2.3. *Procedures for selection model*

Assessing the validity of a model is an important issue in statistical inference. This problem is quite difficult when considering the general sampling scheme. For this case, we provide only a graphical tool. When the shape function is linear and in the balanced case, the increments are independent and identically distributed random variables and so many statistical tests can be used. We present two tests that have been published recently, see [HEN 12] and [VIL 15] (for other tests, the reader can refer to the references in these two papers).

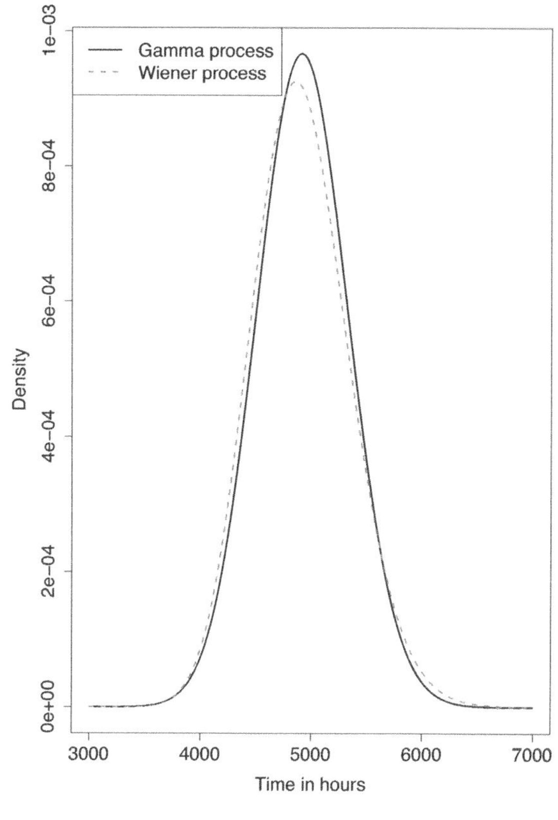

Figure 4.4. *Laser data – comparison of the estimated p.d.f. for time to failure*

A graphical tool: Ye *et al.* [YE 14] have proposed a graphical tool to assess empirically the goodness of fit for the gamma process (see also [BOR 15] for a similar graphical tool for a related degradation model called perturbed gamma process). As a matter of fact, this graphical tool can be used for any degradation model whose increments are independent. Hence, it can be applied to both the Wiener and gamma processes. As discussed earlier in this book, for any $i \in \{1, \ldots, n\}$ and for any $j \in \{1, \ldots, m_i\}$, the increment between $t_{i,j-1}$ and $t_{i,j}$ is denoted by $Y_{i,j}$. Its distribution is assumed to depend on some parameter $\theta_{i,j}$ that can be estimated through any of the previously described methods. The graphical tool relies on the fact that if the data come from a model with c.d.f. $F_{Y_{i,j}}(\cdot;\theta_{i,j})$, then $U_{i,j} := F_{Y_{i,j}}(Y_{i,j};\theta_{i,j})$ is uniformly distributed over $[0;1]$. Replacing $\theta_{i,j}$ by an estimation $\widehat{\theta}_{i,j}$, if using a correct model, the distribution of the random variables $\widehat{U}_{i,j} = F_{Y_{i,j}}(Y_{i,j};\widehat{\theta}_{i,j})$ should hence be closed to the uniform distribution. A QQ plot based on $\widehat{U}_{1,1},\ldots,\widehat{U}_{1,m_1},\ldots,\widehat{U}_{n,1},\ldots,\widehat{U}_{n,m_n}$ is next used to assess (or not) the goodness of fit of the considered degradation model (of course, these random variables are not i.i.d. since they are defined through the estimation of θ). We have applied this graphical tool to the laser data for both the Wiener and gamma processes. The QQ plots are displayed in Figure 4.5. Based on this, it seems that the gamma process fits the data a little better than the Wiener process.

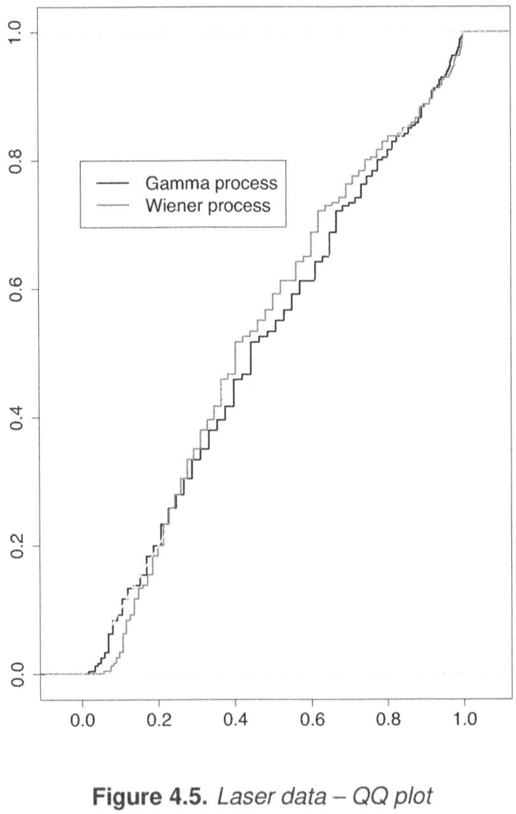

Figure 4.5. *Laser data – QQ plot*

Statistical tests: let us recall that, based on the fact that we are here in the balanced case with a linear shape function, the increments are i.i.d. random variables so that many classical statistical tests can be used. For the Wiener process, there exist many normality tests. We here use the Shapiro–Wilk test [SHA 65]. Similar tests for the gamma distribution are less well known and so we give more details for this case.

Shapiro and Wilk [SHA 65] proposed a goodness-of-fit test for the normal distribution when the parameters are unknown. For the sake of simplicity, we denote by $X_1, \ldots, X_n$ a sample of n i.i.d. random variables and we aim at testing if the common underlying distribution is a normal distribution. The statistic that they proposed is as follows:

$$W = \frac{\left[\sum_{i=1}^{n} a_i X_{(i)}\right]^2}{\sum_{i=1}^{n} (X_i - \bar{X})^2}$$

where $a_1, \ldots, a_n$ are some given coefficients (provided by tables in the literature) and $X_{(1)} < \cdots < X_{(n)}$ are the order statistics. Small values of these statistics lead to reject the null hypothesis.

We now come to the gamma process and we recall two methods from the literature for testing the goodness of fit of the gamma distribution. We still denote by $X_1, \ldots, X_n$ a sample of n i.i.d. random variables and now we aim at testing if the common underlying distribution is a gamma distribution.

Henze *et al.* [HEN 12] have proposed a test based on the empirical Laplace transform of the observations. The key point is that the Laplace transform $\mathcal{L}$ of the gamma distribution $\mathcal{G}(a, b)$ is the only solution of the following ordinary differential equation:

$$(1 + bt)\mathcal{L}'(t) + ab\mathcal{L}(t) = 0 \qquad [4.1]$$

with initial condition $\mathcal{L}(0) = 1$. Considering an estimator $(\widehat{a}_n, \widehat{b}_n)$ of (a, b), they set $Y_i = \widehat{b}_n X_i$ for any $i \in \{1, \ldots, n\}$ and introduce the empirical Laplace transform $\widehat{\mathcal{L}}_n$ of $Y_1, \ldots, Y_n$. Based on [4.1], they next consider statistics of the form:

$$T_n = \int_0^{\infty} Z_n^2(t) w(t) \mathrm{d}t,$$

where w is a continuous positive weight function such that $\int_0^\infty t^4 w(t)\mathrm{d}t < \infty$ and where

$$Z_n(t) = \sqrt{n}\left[(1+t)\widehat{\mathcal{L}}_n'(t) + \widehat{a}_n \widehat{\mathcal{L}}_n(t)\right].$$

The null hypothesis is rejected for large values of T_n. For the weight function, they have considered $w(t) = \exp(-ct)$ and $w(t) = \exp(-ct^2)$ with $c > 0$. For these two choices for the weight function, under the null hypothesis, they have derived an explicit expression of the distribution of the statistics. They have also obtained some asymptotic results.

Villaseñor and González-Estrada [VIL 15] have recently proposed a test based on a ratio of two different estimators of the variance, similarly to the Shapiro–Wilk statistic for testing the normality of a sample. For the sake of simplicity, we now consider the parameterization in terms of the scale parameter β rather than in terms of the rate parameter $b = \beta^{-1}$. The estimators provided by the method of moments for a and β are

$$\hat{a}_n = \frac{\bar{X}_n}{\hat{\beta}_n} = \frac{\bar{X}_n^2}{S_n^2} \quad \text{and} \quad \hat{\beta}_n = \frac{S_n^2}{\bar{X}_n}$$

where

$$\bar{X}_n = \frac{1}{n}\sum_{i=1}^n X_i \quad \text{and} \quad S_n^2 = \frac{1}{n-1}\sum_{i=1}^n (X_i - \bar{X}_n)^2$$

(notice that these estimators are similar to the estimators that could be derived from equation [2.46] in the balanced case, up to the parameterization). Villaseñor and González-Estrada [VIL 15] have provided a new estimator for the parameters, which is based on the following remark: if X is a random variable with gamma distribution $\mathcal{G}(a, b)$ and $Z = \log X$, then $\mathrm{cov}(X, Z) = b^{-1}$. This leads to the following new estimator for β (and thus also for a):

$$\breve{a}_n = \frac{\bar{X}_n}{\breve{\beta}_n} \quad \text{and} \quad \breve{\beta}_n = \frac{1}{n}\sum_{i=1}^n (X_i - \bar{X}_n)(Z_i - \bar{Z}_n)$$

where $Z_i = \log X_i$ for any $i \in \{1, \ldots, n\}$ and $\bar{Z}_n$ is the empirical mean of $Z_1, \ldots, Z_n$. We now have two different estimators of the variance for a gamma distribution, namely S_n^2 and $\breve{\sigma}_n^2 = \bar{X}_n \breve{\beta}_n$. Hence, Villaseñor and González-Estrada have considered the ratio of these estimators: $V_n = S_n^2/\breve{\sigma}_n^2$. Under the null hypothesis, V_n is expected to be close to 1 and they suggest to use the test statistic $V_n^\star = \sqrt{n\breve{a}_n}(V_n - 1)$, which satisfies the following convergence in distribution:

$$V_n^\star \xrightarrow[n\to\infty]{d} \mathcal{N}(0, \rho^2)$$

with $\rho^2 = 1 + a(a\psi_1(a) - 1)$. The asymptotic variance ρ^2 is a non-increasing function of a, with $\lim_{a\to\infty} \rho^2 = 3/2$ and $\lim_{a\to 0} \rho^2 = 2$. It suggests to use quantiles of the $\mathcal{N}(0, 2)$ distribution. The null hypothesis is rejected if $|V_n^\star| > \sqrt{2} z_{1-\epsilon/2}$ where $z_{1-\epsilon/2}$ is the quantile of standard Gaussian distribution of order $1 - \epsilon/2$. This test has been implemented in the `R` package `goft`.

For the laser data, we have applied the Shapiro–Wilk test (Wiener process) and the test proposed by Villaseñor and González-Estrada [VIL 15] (gamma process). We have obtained that $SW = 0.9248$, giving a p-value equal to 1.095×10^{-9}. Hence, for the Wiener process, the null hypothesis is rejected. Besides, we have obtained that $|V_n^\star| = 0.6531$, which is smaller than the critical value 3.92 (for a first-type error equal to 5%). Thus, for the gamma process, the null hypothesis is not rejected. As a conclusion, there is no significant difference between the two degradation models toward some quantity of interest like the average degradation and the time-to-failure distribution. However, both the QQ plots and the statistical tests advocate a better fit of the data by the gamma process rather than by the Wiener process.

4.3. Hudak crack growth data

The average degradation is not linear any more over time for the Hudak crack growth data. Based on the previously discussed models in this book, we consequently propose to use a non-homogeneous gamma process. Looking at the trajectories provided in Figure I.3, we can think of taking either a power shape function, say $A(t) = at^\gamma$, or an exponential shape function, say $A(t) = \exp(at) - 1$. Choosing between these two shape functions is not a trivial task. In some cases, physical considerations can be used to determine which parametric shape function to consider. For instance, a simplification of the Paris–Erdogan law could lead to an exponential shape function, see [LAW 04] for a discussion. To fit the two previously quoted parametric models based on the gamma process, we use the log-likelihood function given by equation [2.42].

Here, we propose to fit the two previously quoted parametric models and also to estimate non-parametrically the shape function for a standard gamma process. In this way, we can compare at least graphically the two parametric models. We have used the approach based on pseudo-likelihood to this purpose (see section 2.8.1). The estimated average degradation is shown in Figure 4.6 for the two parametric models and the non-parametric approach, on the left for the power shape function and on the right for the exponential shape function. It seems that the first function provides a closer result to the non-parametric estimation than the second function. This is confirmed by looking the QQ plot (as discussed for the laser data), see Figure 4.7.

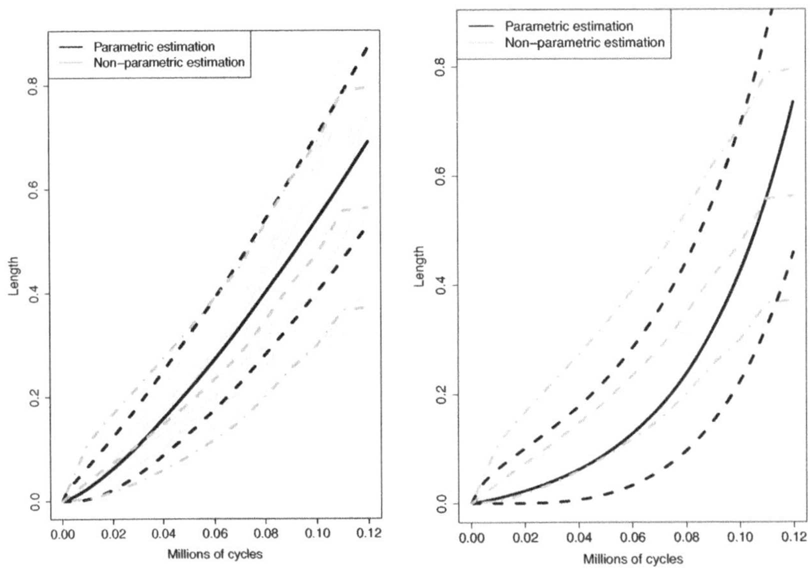

Figure 4.6. *Hudak crack growth data – estimation of the average degradation (left: power shape function; right: exponential shape function)*

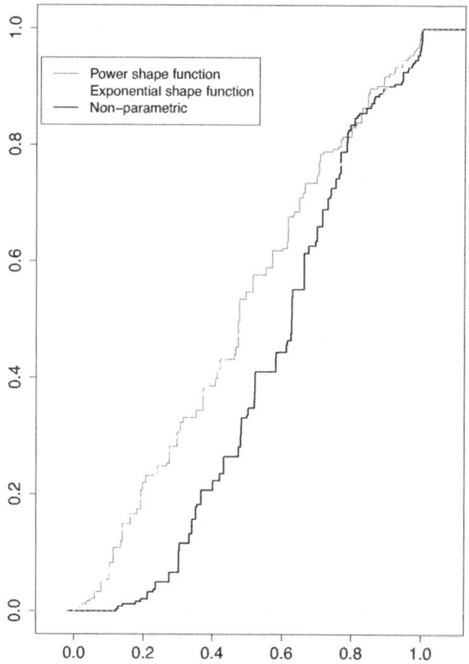

Figure 4.7. *Hudak crack growth data – QQ plot*

Bibliography

[ABD 75] ABDEL-HAMEED M., "A Gamma wear process", *IEEE Transactions on Reliability*, vol. 24, no. 2, pp. 152–153, 1975.

[ABD 14] ABDEL-HAMEED M., *Lévy Processes and their Applications in Reliability and Storage*, Springer, Heidelberg, 2014.

[ABR 72] ABRAMOWITZ M., STEGUN I. (eds.), *Handbook of Mathematical Functions with Formulas, Graphs and Mathematical Tables*, National Bureau of Standards, Applied Mathematics Series 55, 1972.

[AHM 12] AHMAD R., KAMARUDDIN S., "An overview of time-based and condition-based maintenance in industrial application", *Computers and Industrial Engineering Journal*, vol. 63, no. 1, pp. 135–149, 2012.

[AIG 97] AIGNER M., *Combinatorial Theory*, Springer, Berlin, 1997.

[ALB 81] ALBRECHT P., *Dynamische statistische Entscheidungsverfahren für Schadenzahlprozesse*, VVW Karlsruhe, 1981.

[ALM 15] AL MASRY Z., MERCIER S., VERDIER G., "Approximate simulation techniques and distribution of an extended Gamma process", *Methodology and Computing in Applied Probability*, available online, 2015.

[AND 93] ANDERSON P., BORGAN Ø., GILL R. *et al.*, *Statistical Models based on Counting Processes*, Springer, New York, 1993.

[AND 02] ANDERSEN P.K., KEIDING N., "Multi-state models for event history analysis", *Statistical Methods in Medical Research*, vol. 11, pp. 91–115, 2002.

[ASH 90] ASH R.B., *Information Theory*, Dover Publications Inc., New York, 1990.

[ASM 07] ASMUSSEN S., GLYNN P.W., *Stochastic Simulation: Algorithms and Analysis, Stochastic Modeling and Applied Probability*, Springer, New York, vol. 57, 2007.

[AVE 98] AVEN T., JENSEN U., *Stochastic Models in Reliability*, Springer, New York, 1998.

[AVR 03] AVRAMIDIS A.N., LÉCUYER P., TREMBLAY P.-A., "Efficient simulation of Gamma and variance-Gamma processes", *Simulation Conference, Proceedings of the 2003 Winter*, IEEE, vol. 1, pp. 319–326, 2003.

[BAG 04] BAGDONAVICIUS V., NIKULIN M., *Accelerated Life Models, Modeling and Statistical Analysis*, Chapman and Hall/CRC Press, Boca Raton, 2004.

[BAR 65] BARLOW R.E., PROSCHAN F., *Mathematical Theory of Reliability*, vol. 17 of *Classics in Applied Mathematics*, Society for Industrial and Applied Mathematics (SIAM), Philadelphia, PA, 1965, 1996.

[BAR 75] BARLOW R.E., PROSCHAN F., *Statistical Theory of Reliability and Life Testing*, Holt, Rinehart and Winston, Inc., New York, 1975.

[BAR 86] BARNDORFF-NIELSEN O., BLAESILD P., "A note on the calculation of Bartlett adjustments", *Journal of the Royal Statistical Society*, vol. 49, pp. 353–358, 1986.

[BER 96] BERTOIN J., *Lévy processes, Cambridge Tracts in Mathematics*, Cambridge University Press, Cambridge, vol. 121, 1996.

[BIL 95] BILLINGSLEY P., *Probability and Measure*, Wiley Series in Probability and Mathematical Statistics, Third edition, John Wiley & Sons, Inc., New York, 1995.

[BIL 99] BILLINGSLEY P., *Convergence of Probability Measures*, John Wiley & sons, New York, 1999.

[BON 82] BONDESSON L., "On simulation from infinitely divisible distributions", *Journal of Applied Probability and Advances in Applied Probability*, vol. 14, no. 4, pp. 855–869, 1982.

[BOR 15] BORDES L., PAROISSIN C., SALAMI A., "Parametric inference in a perturbed gamma degradation process", *Commun. Statist – Theor. Meth.*, 2015.

[BRE 81] BREMAUD P., *Point Processes and Queues*, Springer, New York, 1981.

[CHH 89] CHHIKARA R.S., LEROY F.J., *The Inverse Gaussian Distribution: Theory, Methodology and Applications*, Dekker New York, 1989.

[ÇIN 77a] ÇINLAR E., "Shock and wear models and Markov additive processes", in TSOKOS C., SHIMI I., (eds.), *The Theory and Applications of Reliability with Emphasis on Bayesian and Nonparametric Methods*, Academic Press, New York, pp. 193–214, 1977.

[ÇIN 77b] ÇINLAR E., BAZANT Z.P., OSMAN E., "Stochastic process for extrapolating concrete creep", *Journal of Engineering Mechanics Division, ASCE*, vol. 103, no. EM6, pp. 1069–1088, 1977.

[ÇIN 80] ÇINLAR E., "On a generalization of gamma processes", *Journal of Applied Probability*, vol. 17, no. 2, pp. 467–480, 1980.

[ÇIN 11] ÇINLAR E., *Probability and stochastics*, *Graduate Texts in Mathematics*, Springer, New York, vol. 261, 2011.

[COM 99] COMMENGES D., "Multi-state Models in Epidemiology", *Lifetime Data Analysis*, vol. 5, pp. 315–327, 1999.

[COR 87] CORDEIRO G., "On the corrections to the likelihood ratio statistics", *Biometrika*, vol. 74, pp. 265–247, 1987.

[COX 55] COX D.R., "Some statistical methods connected with series of events.", *Journal of The Royal Statistical Society Series B-statistical Methodology*, vol. 17, pp. 129–164, 1955.

[COX 87] COX D.R., REID N., "Parameter orthogonality and approximate conditional inference", *Journal of The Royal Statistical Society Series B-statistical Methodology*, vol. 49, pp. 1–39, 1987.

[CRA 69] CRAMER H., "On streams of random events", *Skand. Aktuar Tidskr. Suppl.*, vol. 85, pp. 13–23, 1969.

[DAL 03] DALEY D.J., VERE-JONES D., *An introduction to the theory of point processes - Volume I: general theory and structure*, Springer Science & Business Media, second edition, vol. 1, 2003.

[DOK 92] DOKSUM K.A., HOYLAND A., "Models for variable-stress accelerated life testing experiment based on Wiener processes and the inverse Gaussian distribution", *Technometrics*, vol. 34, pp. 74–82, 1992.

[DOK 95] DOKSUM K.A., NORMAND S.L.T., "Gaussian models for degradation processes – Part I: methods for the analysis of biomarker data", *Lifetime Data Analysis*, vol. 1, pp. 131–144, 1995.

[DOK 96] DOKSUM K.A., NORMAND S.L.T., "Models for degradation processes and event times based on gaussian processes", in JEWELL N.P.E.A., (ed.),, *Lifetime Data: Models in Reliability and Survival Analysis*, Kluwer academic publishers, pp. 85–91, 1996.

[DOO 53] DOOB J.L., *Stochastic Processes*, New York Wiley, vol. 101, 1953.

[DUD 11] DUDLEY R.M., NORVAIA R., *Concrete Functional Calculus*, Springer, New York, 2011.

[DUF 91] DUFRESNE F., GERBER H.U., SHIU E.S.W., "Risk theory with the Gamma process", *ASTIN Bulletin*, vol. 21, no. 2, pp. 177–192, 1991.

[DYK 81] DYKSTRA R., LAUD P., "A Bayesian nonparametric approach to reliability", *Annals of Statistics*, vol. 9, no. 2, pp. 356–367, 1981.

[EBE 09] EBERLEIN E., "Jump-type Lévy Processes", in MIKOSCH T., KREISS J.P., DAVIS R.A., ANDERSEN T.G., (eds.), *Handbook of Financial Time Series*, pp. 439–455, Berlin Heidelberg, Springer, 2009.

[ESA 73] ESARY J.D., MARSHALL A.W., PROSHAN F., "Shock models and wear processes", *Annals of Probability*, vol. 1, pp. 627–649, 1973.

[ESC 06] ESCOBAR L., MEEKER W., "A review of accelerated test models", *Statistical Science*, vol. 21, pp. 522–577, 2006.

[FEN 94] FENG W., ADACHI K., KOWADA M., "Optimal replacement under additive damage in a Poisson random environment", *Communications in Statistics – Stochastic Models*, vol. 10, pp. 679–700, 1994.

[FIN 13] FINKELSTEIN M., CHA J.H., *Stochastic Modeling for Reliability: Shocks, Burn-in and Heterogeneous Populations*, Springer Science & Business Media, 2013.

[GIO 15] GIORGIO M., GUIDA M., PULCINI G., "A new class of Markovian processes for deteriorating units with state dependent increments and covariates", *IEEE Transactions on Reliability*, vol. 64, no. 2, pp. 562–578, 2015.

[GRA 91] GRANDELL J., *Aspects of Risk Theory*, Springer, New York, 1991.

[GRA 97] GRANDELL J., *Mixed Poisson Processes*, Chapman & Hall, London, 1997.

[GRA 07] GRADSHTEYN I., RYZHIK I. (eds.), *Table of Integrals, Series and Products*, Academic Press, Burlington, MA, 2007.

[GUI 12] GUIDA M., POSTIGLIONE F., PULCINI G., "A time-discrete extended Gamma process for time-dependent degradation phenomena", *Reliability Engineering and System Safety*, vol. 105, pp. 73–79, 2012.

[GUT 05] GUT A., HÜSLER J., "Realistic variation of shock models", *Statistics & Probability Letters*, vol. 74, pp. 187–204, 2005.

[HEN 12] HENZE N., MEINTANIS S., EBNE B., "Goodness-of-fit tests for the Gamma distribution based on the empirical Laplace transform", *Communications in Statistics - Theory and Methods*, vol. 41, pp. 1543–1556, 2012.

[HOU 99] HOUGAARD P., "Multi-state models: a review", *Lifetime Data Analysis*, vol. 5, pp. 239–264, 1999.

[HUD 78] HUDAK S., SAXENA A., BUCCI R. *et al.*, Development of standard methods of testing and analyzing fatigue crack growth rate data, Technical report AFML-TR-78-40, Westinghouse R&D Center, Westinghouse Electric Corporation. Pittsburgh, PA, 1978.

[IMA 13] IMAI J., KAWAI R., "Numerical inverse Lévy measure method for infinite shot noise series representation", *Journal of Computational and Applied Mathematics*, vol. 253, no. 0, pp. 264–283, 2013.

[ISH 04] ISHWARAN H., JAMES L.F., "Computational methods for multiplicative intensity models using weighted Gamma processes", *Journal of the American Statistical Association*, vol. 99, no. 465, pp. 175–190, 2004.

[JAC 87] JACOD J., SHIRYAEV A.N., *Limit Theorems for Stochastic Processes*, *Grundlehren der Mathematischen Wissenschaften*, Springer, Berlin, vol. 288, 1987.

[JAR 06] JARDINE A.K., LIN D., BANJEVIC D., "A review on machinery diagnostics and prognostics implementing condition-based maintenance", *Mechanical Systems and Signal Processing*, vol. 20, no. 7, pp. 1483–1510, 2006.

[JOH 95] JOHNSON N., KOTZ S., BALAKRISHNAN N., *Continuous Univariate Distributions*, Wiley-Interscience, New York, vol. 2, 1995.

[JØR 82] JØRGENSEN B., *Statistical Properties of the Generalized Inverse Gaussian Distribution*, Springer-Verlag, New York, 1982.

[KAH 98] KAHLE W., LEHMANN A., "Parameter estimation in damage processes: dependent observations of damage increments and first passage time", in KAHLE W.E.A., (ed.),, *Advances in Stochastic Models for Reliability, Quality and Safety*, Statistics for Industry and Technology, Birkhauser, pp. 139–152, 1998.

[KAH 04] KAHLE W., "Simultaneous confidence regions for the parameters of damage processes", *Statistical Papers*, vol. 35, pp. 27–41, 2004.

[KAH 06] KAHLE W., WENDT H., "Statistical analysis of some parametric degradation models", in NIKULIN M., COMMENGES D., HUBER C., (eds.), *Probability, Statistics and Modeling in Public Health*, Springer Science & Business Media, pp. 266–79, 2006.

[KAH 10a] KAHLE W., LEHMANN A., "The Wiener process as a degradation model: modeling and parameter estimation", in NIKULIN M., LIMNIOS N., BALAKRISHNAN N. *et al.* (eds.), *Advances in Degradation Modeling*, Birkhäuser, pp. 127–146, Boston, 2010.

[KAH 10b] KAHLE W., WENDT H., "Parametric shock models", in NIKULIN M., LIMNIOS N., BALAKRISHNAN N. *et al.* (eds.), *Advances in Degradation Modeling*, pp. 77–104, Birkhäuser, Boston, 2010.

[KLE 03] KLEIN J., MOESCHBERGER M., *Survival Analysis, Techniques for Censored and Truncated Data*, Springer-Verlag, New York, 2003.

[KON 92] KÖNIG D., SCHMIDT V., *Zufällige Punktprozesse*, B.G. Teubner, Stuttgart, 1992.

[KOT 06] KOTZ S., BALAKRISHNAN N., READ B.C. *et al.* (eds.), *Enclyclopedia of Statistical Sciences*, Wiley-Interscience, Hoboken, NJ, vol. 1, 2006.

[LAS 95] LAST G., BRANDT A., *Marked Point Processes on the Real Line*, Springer, New York, 1995.

[LAU 96] LAUD P., SMITH A., DAMIEN P., "Monte Carlo methods for approximating a posterior hazard rate process", *Statistics and Computing*, vol. 6, no. 1, pp. 77–83, Springer, 1996.

[LAW 03] LAWLESS J., *Statistical Models and Methods for Lifetime Data*, Wiley-Interscience, Hoboken, N.J., 2003.

[LAW 04] LAWLESS J., CROWDER M., "Covariates and random effects in a Gamma process model with application to degradation and failure", *Lifetime Data Analysis*, vol. 10, pp. 213–227, 2004.

[LEB 93] LEBLEBICI Y., KANG S., *Hot-Carrier Reliability of MOS VLSI Circuits*, Springer Science & Business Media, LLC, 1993.

[LIN 14] LING M., TSUI K., BALAKRISHNAN N., "Accelerated degradation analysis for the quality of a system based on the Gamma process", *IEEE Transactions on Reliability*, vol. 64, no. 1, pp. 463–472, 2014.

[LU 96] LU C.J., MEEKER W.Q., ESCOBAR L.A., "A comparison of degradation and failure-time analysis methods for estimating a time-to-failure distribution", *Statistics Sinica*, vol. 6, no. 3, pp. 531–546, 1996.

[LU 97] LU J., PARK J., YANG Q., "Statistical inference of a time-to-failure distribution derived from linear degradation data", *Technometrics*, vol. 39, no. 4, pp. 391–400, 1997.

[MAC 08] MACLACHLAN G., KRISHNAN T., *The EM Algorithm and Extensions*, Wiley, Hoboken, NJ, 2008.

[MAR 07] MARSHALL A.W., OLKIN I., *Life Distributions*, Springer, New York, 2007.

[MCC 01] MCCULLOCH C.E., NEUHAUS J.M., *Generalized Linear Mixed Models*, Wiley Online Library, 2001.

[MEE 98] MEEKER W.Q., ESCOBAR L.A., *Statistical Methods for Reliability Data*, J. Wiley & Sons, New York, 1998.

[MIK 98] MIKOSCH T., *Elementary Stochastic Calculus with Finance in View*, World Scientific Publishing Singapore, 1998.

[MOR 66] MOREY R.C., "Some stochastic properties of a compound-renewal damage model", *Operations Research, INFORMS*, vol. 14, no. 5, pp. 902–908, 1966.

[MÖR 10] MÖRTERS P., PERES Y., *Brownian Motion*, Cambridge University Press, Cambridge, UK, 2010.

[NAG 07] NAGAWAKA T., *Shock and Damage Models in Reliability Theory*, Springer, London, 2007.

[NEL 02] NELSON W., *Accelerated Testing, Statistical Models, Test Plans and Data Analysis*, J. Wiley & Sons, Hoboken, N.J., 2002.

[NEV 77] NEVEU J., "Processus ponctuels", in HENNEQUIN P.-L., (ed.), *Ecole d'Eté de Probabilités de Saint-Flour VI-1976, Lecture Notes in Mathematics*, Springer Berlin Heidelberg, vol. 598, pp. 249–445, 1977.

[NIK 10] NIKULIN M., LIMNIOS N., BALAKRISHNAN N. *et al.*, (eds.), *Advances in Degradation Modeling, Applications to Reliability, Survival Analysis and Finance*, Birkhäuser, Boston, 2010.

[PAD 79] PADGETT W.J., WEI L.J., "Estimation for the three parameter inverse Gaussian distribution", *Communications in Statistics - Theory and Methods*, vol. 8, pp. 129–137, 1979.

[PAR 05] PARK C., PADGETT W., "Accelerated degradation models for failure based on geometric Brownian motion and Gamma process", *Lifetime Data Analysis*, vol. 11, pp. 511–527, 2005.

[PAR 14] PAROISSIN C., SALAMI A., "Failure time of non-homogeneous Gamma process", *Communications in Statistics - Theory and Methods*, Taylor & Francis, vol. 43, no. 15, pp. 3148–3161, 2014.

[PAR 15] PAROISSIN C., RABEHASAINA L., "First and last passage times of spectrally positive Lévy processes with application to reliability", *Methodology and Computing in Applied Probability*, vol. 17, no. 2, pp. 351–372, 2015.

[PET 95] PETROV V., *Limit Theorems of Probability Theory – Sums of Independent Random Variables*, Oxford University Press Inc., New York, 1995.

[PUL 13] PULCINI G., "Modeling the mileage accumulation process with random effects", *Communications in Statistics – Theory and Methods*, vol. 42, pp. 2661–2683, 2013.

[ROB 88] ROBERTSON T., WRIGHT F., DYKSTRA R., *Order Restricted Statistical Inference*, Wiley and Sons, Chichester, UK, 1988.

[ROS 96] ROSS S.M., *Stochastic Processes*, Second edition, John Wiley & Sons Inc., New York, 1996.

[ROS 01] ROSIŃSKI J., "Series Representations of Lévy Processes from the Perspective of Point Processes", *Lévy Processes*, Springer, pp. 401–415, 2001.

[ROS 08] ROSIŃSKI J., "Simulation of Lévy processes", *Encyclopedia of Statistics in Quality and Reliability*, Wiley Online Library, 2008.

[RUD 87] RUDIN W., *Real and Complex Analysis*, Tata McGraw-Hill Education, 1987.

[RYK 10] RYKOV V., BALAKRISHNAN N., NIKULIN M. (eds.), *Mathematical and Statistical Models and Methods in Reliability, Applications to Medicine, Finance and Quality Control*, Birkhäuser, Boston, 2010.

[SAT 99] SATO K., *Lévy Processes and Infinitely Divisible Distributions*, *Cambridge Studies in Advanced Mathematics*, Cambridge University Press, Cambridge, vol. 68, 1999.

[SCH 95] SCHRÖTER K., *Verfahren zur Approximation der Gesamtschadenverteilung: Systematisierung, Techniken und Vergleiche*, Karlsruher Reihe, Verlag Versicherungswirtschaft, 1995.

[SHA 65] SHAPIRO S., WILK M., "An analysis of variance test for normality (complete samples)", *Biometrika*, vol. 52, pp. 591–611, 1965.

[SHA 83] SHAKED M., "Wear and damage processes from shock models", *Reliability Theory and Models*, Proc. Symp., Charlotte/N.C., pp. 43–64, 1983.

[SHA 07] SHAKED M., SHANTHIKUMAR J.G., *Stochastic Orders*, Springer, New York, 2007.

[SI 12] SI X., WANG W., HU C. *et al.*, "Remaining useful life estimation based on a nonlinear diffusion degradation process", *IEEE Transactions on Reliability*, vol. 61, pp. 50–67, 2012.

[SOB 87] SOBCZYK K., "Stochastic models for fatigue damage of materials", *Journal of Applied Probability and Advances in Applied Probability*, vol. 19, pp. 652–673, 1987.

[STI 89] STINEBAUGH W.J., HARRUS A., KNOLLE W., "Correlation of G_m degradation of submicrometer MOSFETs with refractive index and mechanical stress of encapsulation materials", *IEEE Electronic Device Letters*, vol. 36, no. 3, pp. 542–547, 1989.

[TAK 83] TAKEDA E., SUZUKI N., "An empirical model for device degradation due to hot-carrier injection", *IEEE Electronic Device Letters*, vol. 4, no. 4, pp. 111–113, 1983.

[TSA 11] TSAI C.-C., TSENG S.-T., BALAKRISHNAN N., "Mis-specification analyses of Gamma and Wiener degradation processes", *Journal of Statistical Planning and Inference*, vol. 141, no. 12, pp. 3725–3735, 2011.

[VAA 98] VAN DER VAART A., *Asymptotic Statistics*, Cambridge University Press, ccc, 1998.

[VAN 09] VAN NOORTWIJK J., "A Survey of the application of Gamma processes in maintenance", *Reliability Engineering and System Safety*, vol. 94, no. 1, pp. 2–21, 2009.

[VIL 15] VILLASEÑOR J., GONZÁLEZ-ESTRADA E., "A variance ratio test of fit for Gamma distributions", *Statistical Probability Letters*, vol. 96, pp. 281–286, 2015.

[VIR 79] VIRKLER D., HILLBERRY B., GOEL P., "The statistical nature of fatigue crack propagation", *Transactions of the ASME – Journal of Engineering Materials and Technology*, vol. 101, pp. 148–153, 1979.

[VOI 96] VOINOV V.G., NIKULIN M.S., *Unbiased Estimators and their Applications: Multivariate Case*, Kluwer Academic Publishers, vol. 2, 1996.

[VOS 08] VOSE D., *Risk Analysis: a Quantitative Guide*, Third edition, John Wiley & Sons, 2008.

[WAN 97] WANG L., PÖTZELBERGER K., "Boundary crossing probability for Brownian motion and general boundaries", *Journal of Applied Probability*, vol. 34, pp. 54–65, 1997.

[WAN 07] WANG L., PÖTZELBERGER K., "Crossing probabilities for diffusion processes with piecewise continuous boundaries", *Methodology and Computing in Applied Probability*, vol. 9, pp. 21–40, 2007.

[WAN 08] WANG X., "A pseudo-likelihood estimation method for nonhomogeneous Gamma process model with random effects", *Statistica Sinica*, vol. 18, pp. 1153–1163, 2008.

[WAN 09a] WANG X., "Nonparametric estimation of the shape function in a Gamma process for degradation data", *Canadian Journal of Statistics*, vol. 37, no. 1, pp. 102–118, 2009.

[WAN 09b] WANG X., "Semiparametric inference of a class of Wiener processes", *Journal of Time Series Analysis*, vol. 30, pp. 179–207, 2009.

[WAN 10] WANG X., "Wiener processes with random effects for degradation data", *Journal of Multivariate Analysis*, vol. 101, no. 2, pp. 340–351, 2010.

[WAN 14] WANG X., JIANG P., GUO B. *et al.*, "Real-time reliability evaluation with a general Wiener-process-based degradation model", *Quality and Reliability Engineering International*, vol. 30, pp. 205–220, 2014.

[WEN 81] WENTZELL A., *A Course in the Theory of Stochastic Processes*, McGraw-Hill International, 1981.

[WEN 98] WENDT H., A model describing damage processes and resulting first passage times, Report, Otto-von-Guericke University Magdeburg, Faculty for Mathematics, 1998.

[WEN 99] WENDT H., Parameterschätzungen für eine Klasse doppelt–stochastischer Poisson Prozesse bei unterschiedlichen Beobachtungsinformationen, PhD thesis, Otto-von-Guericke University, Magdeburg, Germany, 1999.

[WEN 04a] WENDT H., KAHLE W., "On a cumulative damage process and resulting first passage times", *Applied Stochastic Models in Business and Industry*, vol. 20, pp. 17–26, 2004.

[WEN 04b] WENDT H., KAHLE W., "On parameter estimation for a position-dependent marking of a doubly stochastic Poisson process", in BALAKRISHNAN N., NIKULIN M.S., MESBAH M. *et al.*, (eds.), *Parametric and Semiparametric Models with Applications to Reliability, Survival Analysis and Quality of Life*, Statistics for Industry and Technology, Birkhauser, pp. 473–486, 2004.

[WHI 95] WHITMORE G.A., "Estimating degradation by a Wiener difussion process subject to measurement error", *Lifetime Data Analysis*, vol. 1, pp. 307–319, 1995.

[WHI 97] WHITMORE G.A., SCHENKELBERG F., "Modeling accelerated degradation data using Wiener difussion with a time-scale transformation", *Lifetime Data Analysis*, vol. 3, pp. 27–45, 1997.

[WHI 98] WHITMORE G.A., CROWDER M.J., LAWLESS J.F., "Failure inference from a marker process based on a bivariate Wiener model", *Lifetime Data Analysis*, vol. 4, pp. 229–251, 1998.

[YE 14] YE Z.-S., XIE M., TANG L.-C. *et al.*, "Semiparametric estimation of Gamma processes for deteriorating products", *Technometrics*, vol. 56, no. 4, pp. 504–513, 2014.

[YE 15] YE Z.-S., XIE M., "Stochastic modeling and analysis of degradation for highly reliable products", *Applied Stochastic Models in Business and Industry*, vol. 31, no. 1, pp. 16–32, 2015.

Index

A

accelerating factor, 145
- Arrhenius law, 146
- exponential law, 146
- inverse-linear law, 146
- inverse-log law, 146
- inverse-power law, 146
- power law, 146

asymptotic normality
- gamma process, 117, 120, 122
- Wiener process, 24, 26, 30

B, C, D

Brownian motion, 3
- with nonlinear drift, 46

covariates, 145
Cox process, 152
di-gamma function, 101
distribution
- beta, 51, 84, 136
- Delaporte, 154
- Fisher, 136
- gamma, 49, 153
- Gaussian, 1
- inverse gamma, 135
- inverse Gaussian, 16, 18, 154
- normal, 1
- of a random measure, 58
- Poisson, 54
- uniform, 152

E, F

EM algorithm, 130, 132, 143
exponential integral function, 77, 100
first passage time
- Brownian motion, 14
- Brownian motion with nonlinear drift, 46
- doubly stochastic marked Poisson process, 151, 163
- gamma process, 97, 102, 125
- gamma process with random effects, 136
- time-scaled Wiener process, 46
- Wiener process, 12, 16

frailty, 152
Frullani integral, 72

G

gamma
- distribution, 49
- function, 49
- process, 49, 68

Gaussian vector, 1
generalized hyper-geometric function, 101
greatest convex minorant, 128

H, I

hitting time
- Brownian motion, 14
- Brownian motion with nonlinear drift, 46
- doubly stochastic marked Poisson process, 151, 163
- gamma process, 97, 102, 125
- gamma process with random effects, 136
- time-scaled Wiener process, 46
- Wiener process, 12, 16

increasing failure rate, 19, 107
increments
- independent, 3
- stationary, 3

infinitely divisible, 3, 69
intensity measure, 59

L

Lévy
- measure, 77
- process, 55, 69, 76

Laplace
- functional of a Poisson random measure, 64
- functional of a random measure, 63
- transform of a gamma distribution, 50
- transform of a random variable, 50

likelihood function
- gamma process, 114, 116, 125, 126
- gamma process with random effetcs, 137
- Wiener process, 23

log-likelihood function
- gamma process, 114, 116, 127
- gamma process with random effetcs, 137

lower incomplete gamma function, 101

M

marked point process, 150
maximum likelihood estimator
- Wiener process, 24

mean time to failure
- gamma process, 102
- Wiener process, 19

measure
- atomic, 56
- counting, 57
- diffuse, 57
- intensity, 59
- purely atomic, 57
- random, 57

moments method estimator
- doubly stochastic marked Poisson process, 179
- gamma process, 118, 121
- Wiener process, 30

N, O

new better than used (NBU), 107
order
- hazard rate order, 53, 70
- likelihood ratio order, 53, 70
- reversed hazard rate order, 53, 70
- usual stochastic order, 53, 70

overshoot, 97, 98

P

partial pseudo log-likelihood function, 127
pochammer symbol, 101
Poisson
- compound, 88, 159
- definition, 55
- distribution, 54
- doubly stochastic, 152
- homogeneous, 55
- mixed, 152
- non-homogeneous, 55, 152
- process, 60
- random measure, 54, 59, 60, 64, 68, 73, 77, 78
- simulation, 55

pool adjacent violators algorithm, 128, 139

position-dependent marking, 157
potential function, 99
process
 additive, 46, 55, 69
 counting, 55
 Cox, 152
 gamma, 68
 Gaussian, 3
 Lévy, 55, 69, 76
 marked point, 150
 Poisson, 55
 time-scaled Wiener process, 45
 Wiener, 10
profile pseudo log-likelihood function, 139
pseudo
 likelihood function, 138
 log-likelihood function, 127, 139

R, S

reflection property, 4
remaining lifetime, 106
 Wiener process, 20
self-similarity, 4
simulation
 Brownian bridge sampling, 8
 gamma bridge sampling (GBS), 84
 gamma process, 81, 84, 87
 Karhunen-Loève approximation, 9
 Poisson process, 55
 Poisson random measure, 68
 random walk approximation (RWA), 6, 81
stochastic intensity, 158

T, U

time-to-failure
 Brownian motion, 14
 Brownian motion with nonlinear drift, 46
 doubly stochastic marked Poisson process, 151, 163
 gamma process, 97, 102, 124, 125
 gamma process with random effects, 136
 time-scaled Wiener process, 46
 Wiener process, 12, 16
tri-gamma function, 117
upper incomplete gamma function, 101

Other titles from

iSTE

in

Mathematics and Statistics

2016

CELANT Giorgio, BRONIATOWSKI Michel
Interpolation and Extrapolation Optimal Designs 1

KERN Michel
Numerical Methods for Inverse Problems

2015

DE SAPORTA Benoîte, DUFOUR François, ZHANG Huilong
Numerical Methods for Simulation and Optimization of Piecewise Deterministic Markov Processes

DEVOLDER Pierre, JANSSEN Jacques, MANCA Raimondo
Basic Stochastic Processes

LE GAT Yves
Recurrent Event Modeling Based on the Yule Process
(Mathematical Models and Methods in Reliability Set – Volume 2)

2014

COOKE Roger M., NIEBOER Daan, MISIEWICZ Jolanta
Fat-tailed Distributions: Data, Diagnostics and Dependence
(Mathematical Models and Methods in Reliability Set – Volume 1)

MACKEVIČIUS Vigirdas
Integral and Measure: From Rather Simple to Rather Complex

PASCHOS Vangelis Th
Combinatorial Optimization – 3-volume series – 2nd edition
Concepts of Combinatorial Optimization / Concepts and Fundamentals – volume 1
Paradigms of Combinatorial Optimization – volume 2
Applications of Combinatorial Optimization – volume 3

2013

COUALLIER Vincent, GERVILLE-RÉACHE Léo, HUBER Catherine, LIMNIOS Nikolaos, MESBAH Mounir
Statistical Models and Methods for Reliability and Survival Analysis

JANSSEN Jacques, MANCA Oronzio, MANCA Raimondo
Applied Diffusion Processes from Engineering to Finance

SERICOLA Bruno
Markov Chains: Theory, Algorithms and Applications

2012

BOSQ Denis
Mathematical Statistics and Stochastic Processes

CHRISTENSEN Karl Bang, KREINER Svend, MESBAH Mounir
Rasch Models in Health

DEVOLDER Pierre, JANSSEN Jacques, MANCA Raimondo
Stochastic Methods for Pension Funds

2011

MACKEVIČIUS Vigirdas
Introduction to Stochastic Analysis: Integrals and Differential Equations

MAHJOUB Ridha
Recent Progress in Combinatorial Optimization – ISCO2010

RAYNAUD Hervé, ARROW Kenneth
Managerial Logic

2010

BAGDONAVIČIUS Vilijandas, KRUOPIS Julius, NIKULIN Mikhail
Nonparametric Tests for Censored Data

BAGDONAVIČIUS Vilijandas, KRUOPIS Julius, NIKULIN Mikhail
Nonparametric Tests for Complete Data

IOSIFESCU Marius *et al.*
Introduction to Stochastic Models

VASSILIOU PCG
Discrete-time Asset Pricing Models in Applied Stochastic Finance

2008

ANISIMOV Vladimir
Switching Processes in Queuing Models

FICHE Georges, HÉBUTERNE Gérard
Mathematics for Engineers

HUBER Catherine, LIMNIOS Nikolaos *et al.*
Mathematical Methods in Survival Analysis, Reliability and Quality of Life

JANSSEN Jacques, MANCA Raimondo, VOLPE Ernesto
Mathematical Finance

2007

HARLAMOV Boris
Continuous Semi-Markov Processes

2006

CLERC Maurice
Particle Swarm Optimization

Printed and bound by CPI Group (UK) Ltd, Croydon, CR0 4YY
13/09/2026
14961021-0002